Springer Theses

Recognizing Outstanding Ph.D. Research

Aims and Scope

The series "Springer Theses" brings together a selection of the very best Ph.D. theses from around the world and across the physical sciences. Nominated and endorsed by two recognized specialists, each published volume has been selected for its scientific excellence and the high impact of its contents for the pertinent field of research. For greater accessibility to non-specialists, the published versions include an extended introduction, as well as a foreword by the student's supervisor explaining the special relevance of the work for the field. As a whole, the series will provide a valuable resource both for newcomers to the research fields described, and for other scientists seeking detailed background information on special questions. Finally, it provides an accredited documentation of the valuable contributions made by today's younger generation of scientists.

Theses are accepted into the series by invited nomination only and must fulfill all of the following criteria

- They must be written in good English.
- The topic should fall within the confines of Chemistry, Physics, Earth Sciences, Engineering and related interdisciplinary fields such as Materials, Nanoscience, Chemical Engineering, Complex Systems and Biophysics.
- The work reported in the thesis must represent a significant scientific advance.
- If the thesis includes previously published material, permission to reproduce this must be gained from the respective copyright holder.
- They must have been examined and passed during the 12 months prior to nomination.
- Each thesis should include a foreword by the supervisor outlining the significance of its content.
- The theses should have a clearly defined structure including an introduction accessible to scientists not expert in that particular field.

More information about this series at http://www.springer.com/series/8790

Diana Iruretagoyena Ferrer

Supported Layered Double Hydroxides as CO_2 Adsorbents for Sorption-enhanced H_2 Production

Doctoral Thesis accepted by
Imperial College London, UK

Springer

Author
Dr. Diana Iruretagoyena Ferrer
Department of Chemical Engineering
Imperial College London
London
UK

Supervisor
Prof. David Chadwick
Department of Chemical Engineering
Imperial College London
London
UK

ISSN 2190-5053 ISSN 2190-5061 (electronic)
Springer Theses
ISBN 978-3-319-41275-7 ISBN 978-3-319-41276-4 (eBook)
DOI 10.1007/978-3-319-41276-4

Library of Congress Control Number: 2016943808

Printed on acid-free paper

This Springer imprint is published by Springer Nature
The registered company is Springer International Publishing AG Switzerland

To Raul, Mari and Juan

Supervisor's Foreword

It is every supervisor's wish to be able to write the foreword to an outstanding thesis produced by one's own student. That represents progress—in education, in technology—and I hope this is a fitting tribute to the effort and dedication that has gone into the work. Diana came to Imperial from UNAM, Mexico, already with a strong motivation to do research. As a dedicated engineer, she had a keen interest and enthusiasm in the search for solutions to real problems. Of course, humanity faces many such problems, some of global proportions. Foremost among these is that of climate change. Dealing with this problem is possibly one of our greatest challenges and one in which engineers will certainly play a crucial role. Diana's thesis is her contribution towards that goal.

Diana was attracted from the beginning to the concept of combining adsorption processes with catalytic reactions to enhance conversions where equilibrium normally limits productivity. Such sorption-enhanced processes have widespread potential application, but none greater than in the production of hydrogen. Hydrogen is manufactured and used on a truly large scale today, but it could also hold the key to a carbon-free future if the processes of manufacture can be decarbonised at source. Sorption-enhanced hydrogen production in which carbon dioxide is removed as it is produced can make a major contribution to achieving this and helping to mitigate climate change. The challenge is to find adsorber materials which will operate in the temperature window and reaction environment for such a process.

Diana's approach was to identify hydrotalcite as a promising material for the carbon dioxide adsorber and to recognise that as a layer material, it would be compatible with graphene oxide. Hybrid materials based on hydrotalcites and graphene oxide would therefore have appeal and perhaps yield superior performance in stability and perhaps capacity. Diana synthesised a range of hybrid materials. She characterised them in great detail and measured their adsorption properties. She was able to demonstrate the superior stability under thermal cycling that even incorporation of small quantities of graphene oxide provides, making the novel hybrid materials volume-efficient adsorbers. In the real process, the hybrid materials would operate in the presence of water. Diana constructed a dynamic flow equipment to explore the adsorption performance of the new hybrid materials in the presence of water. She surveyed catalysts that would be suitable for use with the adsorber in the eventual process and tested their selectivity to hydrogen production in the shift reaction. Adding methanol might improve the sorption-enhanced shift process by controlling temperature and generation of extra hydrogen, and Diana make thermodynamic analysis to explore the conditions under which such new process would operate.

Diana's approach and thoroughness are an example to all budding engineering Ph.D. students. Her dedication and determination paid off, and her thesis is a tour de force. It was indeed a pleasure to supervise her.

London, UK
May 2016

Prof. David Chadwick

Abstract

Sorption-enhanced hydrogen production is considered an attractive technology to improve the efficiency of the water gas shift reaction (WGS), which is a key stage in H_2 production by methane steam reforming and coal/biomass gasification. This requires the development of selective catalysts and CO_2 adsorbents that are sufficiently stable to tolerate cyclic regeneration during operation. The present work focuses on the assessment of the adsorption performance of novel layered double hydroxides (hydrotalcites) supported on multiwalled carbon nanotubes and graphene oxide for application in sorption-enhanced processes. Emphasis is placed on the stability and capacity of hybrids prepared with relatively low-carbon support loadings in order to maximise the CO_2 uptake per total volume of adsorbent. This parameter is crucial in the industrial implementation of the technology as it dictates the size of the adsorption units and reactors required.

The coprecipitation of magnesium and aluminium nitrates onto well-dispersed carbon nanotubes or graphene oxide is shown to be an effective preparation method that ensures an adequate interaction between the hydrotalcite and the support. Dispersion of the layered double hydroxides onto the carbon nanostructures produces an increase in the intrinsic adsorption capacity, i.e. CO_2 uptake per mass of hydrotalcite. Compared to other supports such as alumina or carbon nanofibers, hydrotalcite platelets can be stabilised with lower loadings of nanotubes or graphene oxide, and therefore, higher CO_2 capacities per volume of hybrid can be achieved. The mass efficiency of the stabiliser is found to be particularly high in the case of graphene oxide, possibly due to a better charge and geometric compatibility with the oxide layers.

Graphene oxide was found not to modify significantly the number or chemistry of the CO_2 adsorption sites of hydrotalcites but helps to maintain the surface heterogeneity which is otherwise lost during temperature-activated regeneration. The CO_2 isotherm of the fresh adsorbent can be modelled by the Freundlich equation but tends towards an ideal Langmuirian surface upon cycling. The changes in the heterogeneity and the capacity of the surface can be described by the three-parameter Toth isotherm. The presence of alkali was found to increase the adsorption capacity of unsupported and supported layered double hydroxides without influencing the multicycle stability of the materials. Co-adsorption of water was found to enhance the stability of the hybrids and increase the uptake of CO_2 as revealed by breakthrough experiments carried out in a fixed-bed adsorption column. There is no evidence of gasification of the carbon support during operation in the presence of water. Under dry and wet conditions, supported and unsupported materials show fast adsorption kinetics which can be approximated by the linear driving force model.

A thorough thermodynamic analysis demonstrates that besides the widely reported increase in hydrogen production achieved by the sorption enhancement of water gas shift, the in situ removal of CO_2 when water gas shift and methanol decomposition (methanol-to-shift) are conducted simultaneously is an attractive alternative to lift the H_2 yield, while autothermal conditions are procured. The temperature range in which sorption enhancement of water gas shift or methanol-to-shift is thermodynamically convenient (573–773 K) coincides with the optimum operation window of the carbon supported hydrotalcites developed in this research. A preliminary catalyst screening shows that $Cu/ZnO/Al_2O_3$ and Pt/CeO_2 are promising candidates to catalyse water gas shift and methanol decomposition.

List of Publications:

Articles

D. Iruretagoyena, M. Shaffer and D. Chadwick, Influence of Alkali Metals (Na, K, Cs) on CO_2 Adsorption by Layered Oxides Supported on Graphene Oxide, *Ind. Eng. Chem. Res.* 54 (2015), 11610.

D. Iruretagoyena, M. Shaffer and D. Chadwick, Layered Double Oxides Supported on Graphene Oxide for CO_2 Adsorption: Effect of Support and Residual Sodium, *Ind. Eng. Chem. Res.* 54 (2015), 6781.

D. Iruretagoyena, M. Shaffer and D. Chadwick, Adsorption of carbon dioxide on graphene oxide supported layered double oxides, *Adsorption.* 20 (2014), 321.

A. Garcia-Gallastegui, **D. Iruretagoyena**, V. Gouvea, M. Mokhtar, A. M. Asiri, S. N. Basahel, S. A. Al-Thabaiti, A. O. Alyoubi, D. Chadwick and M. S. P. Shaffer, Graphene oxide as support for layered double hydroxides: Enhancing the CO_2 sorption capacity, *Chem. Mater.* 24 (2012), 4531.

A. Garcia-Gallastegui, **D. Iruretagoyena**, M. Mokhtar, A. M. Asiri, S. N. Basahel, S. A. Al-Thabaiti, A. O. Alyoubi, D. Chadwick and M. S. P. Shaffer, Layered double hydroxides supported on multi-walled carbon nanotubes: preparation and CO_2 adsorption characteristics, *J. Mater. Chem.* 22 (2012), 13932. ***Cover of the Journal.***

Conferences

D. Iruretagoyena, X. Huang, M. Shaffer, K. Hellgardt and D. Chadwick, Graphene oxide supported hydrotalcites for pre-combustion CO_2 capture, **oral presentation**, ISCRE 23, Bangkok, Thailand, September 2014.

D. Iruretagoyena, A. Garcia-Gallastegui, K. Hellgardt, M. Shaffer and D. Chadwick, Adsorption of carbon dioxide on graphene oxide supported hydrotalcites, **oral presentation (plenary session)**, FOA11, Baltimore, MD, USA, May 2013. *Travel grant awarded by conference organisers.*

D. Iruretagoyena, A. Garcia-Gallastegui, K. Hellgardt, M. Shaffer and D. Chadwick, Supported hydrotalcites for sorption enhanced water gas shift reaction, poster presentation, ISCRE 22, Maastricht, Netherlands, September 2012. ***Prize for best poster.***

A. Garcia-Gallastegui, **D. Iruretagoyena**, M. Mokhtar, A. M. Asiri, S. N. Basahel, S. A. Al-Thabaiti, A. O. Alyoubi, D. Chadwick and M. S. P. Shaffer, Graphene oxide/Carbon nanotube supported layered double hydroxides for CO_2 capture applications, poster presentation, ChemOnTubes, France, April 2012.

A. Garcia-Gallastegui, **D. Iruretagoyena**, M. Mokhtar, A. M. Asiri, S. N. Basahel, S. A. Al-Thabaiti, A. O. Alyoubi, D. Chadwick and M. S. P. Shaffer, Graphene oxide supported layered double hydroxides for CO_2 capture applications, poster presentation, Graphene, Brussels, April 2012.

D. Iruretagoyena, K. Hellgardt and D. Chadwick, Sorption enhanced water gas shift. Adsorption of CO_2 on carbon supported hydrotalcites, **seminar**, DICP, Dalian, China, July 2014.

Acknowledgements

I would like to express my deep gratitude to Prof. David Chadwick for his guidance, enthusiastic encouragement, and useful advice throughout my research project, for his continuous support and interest in my professional career. It has been a privilege to work and learn from you. Thanks for all the nice group trips to the countryside and for all the stories that made me feel happy and motivated during my Ph.D.

I am very grateful to Prof. Klaus Hellgardt for his valuable advice and support and for the organisation of interesting group meetings. I would also like to acknowledge Prof. Milo Shaffer and Dr. Ainara Garcia-Gallastegui for their important contributions to this work and for the overall fruitful collaboration established. Thanks to Dr. Esat Alpay for his advice in aspects related to my work and Andrew Macey and Patricia Carry for their assistance.

Many thanks to all my colleagues from David's and Klaus's groups: Sami, Laura, Dazhi, Nicolas, Kai, Saeed, Chen, Reem, Sumana, Essam, Carmen, Helena, Jane, Xiaowen, Maha, Pedro Arcelus, and Tazli. Special thanks to my friends Rafa, Judith, Julio, Rajesh, Ana Gil, Pedro Bastos, Ana Narvaez, Andres, Zara, John, and Lucia for all the happy times that we have spent together including the incredible Olympic Games. Thanks to all my friends from Mexico for being always there for me.

I acknowledge my sponsors: Consejo Nacional de Ciencia y Tecnologia (CONACyT), Secretaria de Educacion Publica (SEP), and Imperial College London.

Finally, I would like to express my gratitude to the most important people in my life:

I am deeply grateful to Raul for all his love and continuous support, for being the best company that I could possibly have during this journey and for always being there to motivate me and for making me feel happy every day. I am very thankful for all his advice, assistance, and valuable contributions to this research.

I would like to thank my parents Mari and Juan for their unconditional love and for being extremely supportive with me. This achievement is also yours. I love you both immensely.

Thanks to all the rest of my family. I am especially grateful to Lilia, Armida, and Ana for their constant encouragement and for the nice weekly conversations.

Declaration of Originality

I hereby declare that this thesis is the result of my own work, and that any ideas or quotations from the work of other people are appropriately referenced.[1]

London, UK
2014

Diana Iruretagoyena Ferrer

Contents

Nomenclature

a	Exterior surface area of the pellet per volume of bed (m^2 m^{-3})
b	Gas–solid interaction parameter for Langmuir and Toth models (bar^{-1})
b_0	Affinity constant
Bi_m	Biot mass transfer number (–)
c_i	Molar concentration of species i (mol m^{-3})
c_i^o	Inlet molar concentration of species i (mol m^{-3})
C_p	Molar specific heat capacity at constant pressure (J mol^{-1} K^{-1})
C_v	Molar specific heat capacity at constant volume (J mol^{-1} K^{-1})
D	Overall diffusion coefficient of the adsorptive through the pore system (m^2 s^{-1})
D_e	Effective pore diffusion coefficient (m^2 s^{-1})
D_k	Knudsen diffusion coefficient (m^2 s^{-1})
D_m	Gas-phase molecular diffusion coefficient (m^2 s^{-1})
D_p	Intraparticle diffusion coefficient (m^2 s^{-1})
D_z	Axial dispersion coefficient (m^2 s^{-1})
d_p	Equivalent diameter of the particles (m)
d_{pore}	Pore diameter (nm)
D_s	Surface diffusion coefficient (m^2 s^{-1})
d_t	Outer tube diameter (m)
$E(t)$	Residence time distribution function (–)
H	Henry's law constant for adsorption equilibrium (mol kg^{-1} bar^{-1})
I_0	Zeroth-order modified Bessel function of the first type (–)
k	Freundlich gas–solid interaction parameter (mol kg^{-1})
k_{ads}	Adsorption rate coefficient (bar^{-1} s^{-1})
K_D	Ergun equation coefficient for dynamic (inertial) losses (–)
k_{des}	Desorption rate coefficient (s^{-1})
k_g	Thermal conductivity of the gas mixture (J m^{-1} s^{-1} K^{-1})
k_m	Effective interparticle mass transfer coefficient (m s^{-1})

k_{LDF}	Effective intraparticle mass transfer coefficient (s^{-1})
k_p	Thermal conductivity of the particles (J m^{-1} s^{-1} K^{-1})
K_V	Ergun equation coefficient for viscous losses (–)
k_z	Effective thermal conductivity of the packed bed in axial direction (J m^{-1} s^{-1} K^{-1})
L	Length of the tube (m)
L_{bed}	Length of the packed bed (m)
L_C	Characteristic length (m)
L_{column}	Total length of the adsorption column (m)
m	Monolayer capacity (mol kg^{-1})
m_a	Mass of the adsorbent recovered from the adsorption column (g)
m/z	Mass-to-charge ratio of fragments in mass spectrometry (–)
n	Exponent constant in Freundlich equation (–)
n_i	Molar flow rate of species i (mol s^{-1})
n_i^o	Inlet molar flow rate of species i (mol s^{-1})
P	Total pressure of the system (bar)
P_b	Pressure of the adsorption column (bar)
p_i	Partial pressure of species i (bar)
P_o	Saturation pressure at a given temperature (bar)
R	Universal gas constant (J mol^{-1} K^{-1} or equivalent)
Q_{ads}	Isosteric heat of adsorption (J mol^{-1})
Q_{Ar-bk}	Volumetric flow rate of the argon blank (m^3 s^{-1})
Q_f	Feed volumetric flow rate of the gas through the adsorption column (m^3 s^{-1})
q_i	Amount of species i in the adsorbed phase (mol kg^{-1})
q_i^*	Amount of species i in the adsorbed phase that are in equilibrium with the bulk-phase concentration (mol kg^{-1})
q_o	Initial amount of species i in the adsorbed phase (mol kg^{-1})
r	Equivalent radius of particles (m)
Re	Reynolds number (–)
r_{hybrid}	Thermogravimetric residue of the hybrid (mg)
r_{LDH}	Thermogravimetric residue of the pure LDH (mg)
$r_{support}$	Thermogravimetric residue of the support (mg)
S	Selectivity of reactions towards a specific product (–)
Sc	Schmidt number (–)
Sh	Sherwood number (–)
S_i	Entropy of phase i (J mol^{-1} K^{-1})
s^3	Third moment of residence time distribution (s^3)
T	Temperature (K)
t	Time (s) or exponent constant in Toth equation (–)
T_b	Temperature of the adsorption column (K)
t_m	Residence time distribution mean (s)
t_o	Time when the step input occurs (s)

t_{st}	Stoichiometric time (s)
T_w	Temperature of the wall of the column (K)
t_∞	Time necessary to measure the gas inlet concentration after breakthrough (s)
U	Overall heat transfer coefficient (J m^{-2} s^{-1} K^{-1})
u	Superficial velocity (m s^{-1})
v	Interstitial velocity (m s^{-1})
V_b	Volume of the bed of adsorbent (m^3)
V_{CST}	Volume of ideal continuous stirred tank in the dead volume (m^3)
V_{PF}	Volume of ideal plug flow in the dead volume (m^3)
v_{pore}	Pore volume (m^3 g^{-1})
X	Conversion (–)
Y	Yield of reactions towards a specific product (–)
y_i	Molar fraction of species *i* in the gas phase (–)
z	Position in the axial direction (m)

Greek Symbols

α	Ratio of ideal continuous stirred tank volume to total dead volume (–)
α_i	Fraction of type *i* domains in Freundlich formalism (–)
$\alpha_{i/Ar}$	Proportionality factor for the molar ratio of species *i* to argon in mass spectrometry (–)
β	Heating rate in temperature-programmed experiments (K s^{-1})
β_{CO_2}	Quotient of detected mass-to-charge ratios 44 and 28 (–)
β_{CH_3OH}	Quotient of detected mass-to-charge ratios 31 and 15 (–)
ΔG_{ads}	Free energy of adsorption (J mol^{-1})
ΔH_{ads}	Enthalpy of adsorption (J mol^{-1})
ΔS_{ads}	Entropy of adsorption (J mol^{-1})
ΔT	Temperature difference (K)
$\varepsilon, \varepsilon_b$	Void fraction of the packing (–)
ε_p	Porosity of the particles (–)
ε_T	Total intraparticle and interparticle voidage (–)
θ	Fractional area of adsorbent covered by adsorbate (–)
μ_{gas}	Dynamic viscosity of the gas (kg m^{-1} s^{-1})
υ_{ij}	Stoichiometric coefficient of species *i* in reaction *j* (–)
ξ_j	Extent of reaction *j* (mol s^{-1})
ρ_b	Density of the packed bed (kg m^{-3})
ρ_p	Density of the particles (kg m^{-3})
$\rho_{skeletal}$	Skeletal density of the solid (kg m^{-3})
σ^2	Residence time distribution variance (s^2)
τ	Tortuosity (–)

Acronyms

AES	Atomic emission spectroscopy
AR	Autothermal reforming
BET	Brunauer–Emmett–Teller model of physical adsorption
BJH	Barrett–Joyner–Halenda method for determination of pore sizes and distributions
CCS	Carbon capture and storage
CNF	Carbon nanofibre
CST	Continuous stirred tank
EDS	Energy-dispersive spectroscopy
FWHM	Full width at half maximum
GC	Gas chromatography
GO	Graphene oxide
HT	Hydrotalcite
HRTEM	High-resolution transmission electron microscopy
ICP	Inductively coupled plasma spectroscopy
ID	Internal diameter
IGCC	Integrated gasification combined cycle
IRS	Infrared spectroscopy
LDF	Linear driving force
LDH	Layered double hydroxide
LDO	Layered double oxide
MS	Mass spectrometry
MTS	Methanol-to-shift
MTZ	Mass transfer zone
MWCNT	Multiwalled carbon nanotube
NO_x	Generic expression for nitrogen oxides
OES	Optical emission spectroscopy
PF	Plug flow
PO	Partial oxidation
PSA	Pressure-swing adsorption
RTD	Residence time distribution
SEM	Scanning electron microscopy
SEMTS	Sorption-enhanced methanol-to-shift
SESMR	Sorption-enhanced methane steam reforming
SEWGS	Sorption-enhanced water gas shift
SOx	Generic expression for sulphur oxides
SR	Steam reforming
SWCNT	Single-walled carbon nanotube
TCD	Thermal conductivity detector
TEM	Transmission electron microscopy
TGA	Thermogravimetric analysis
TPD	Temperature-programmed desorption

TPR	Temperature-programmed reduction
TSA	Temperature swing adsorption
WGS	Water gas shift
XRD	X-ray diffraction
ZDC	Zirconia-doped ceria

List of Figures

List of Tables

Chapter 1
Introduction

1.1 Research Motivation

Since the beginning of the Industrial Revolution, the prosperity of societies has relied on the use of fossil fuels (coal, oil and natural gas) as a source of energy. With the global reserves of easily-accessible fossil resources constantly dropping and the energy demand increasing, there is a latent risk of a major shortage that would drive fuel prices up to critical levels. Additionally, the use of fossil fuels is associated with the accumulation of carbon dioxide and other greenhouse gases in the atmosphere. Increasingly stringent regulations on the permissible level of emission of NOx, SOx and CO_2 makes it necessary to invest large amounts of hydrogen and energy on cleaning processes exacerbating the escalation of fuel prices. Nowadays between 15 and 20 % of the energy contained in an oil barrel is dedicated to produce environmentally acceptable transportation fuels [1].

In this context, technologies that improve the energy efficiency and render environmentally friendlier processes are of great importance in industry. Hydrogen production is receiving significant attention due to its crucial role as feedstock in chemical industries and refineries and to its application as an ultraclean fuel. Nowadays, most of the worldwide supply (more than 12 trillion cubic feet per year) [2] is consumed in the production of ammonia and methanol or used in refineries to clean and upgrade crude oils by hydrotreating. Additionally, H_2 can be used in electrochemical cells or in combustion engines with the advantage of eliminating pollutant emissions by end users.

Hydrogen can be obtained from fossil fuels, biofuels, alcohols or water (electrolysis) through heterogeneously catalysed reactions. The competitiveness of the different routes depends on the economics of the process which are determined by the efficiency of the catalyst, the scale of the production and the cost of the downstream separation units. H_2 usually coexists with other gaseous products, mainly carbon dioxide, and needs to be purified before it is used. Consequently, the enhancement of hydrogen yield, the simplification of its purification and the

D. Iruretagoyena Ferrer, *Supported Layered Double Hydroxides as CO_2 Adsorbents for Sorption-enhanced H_2 Production*, Springer Theses,
DOI 10.1007/978-3-319-41276-4_1

reduction of the energy demand are crucial considerations in the development of new technologies for the generation of H_2.

Sorption-assisted hydrogen production stands out as one of the most promising strategies to reduce the complexity of the industrial processes while the energy efficiency is markedly increased [3–5]. It consists of the in situ sequestration of CO_2 by adsorption as it is formed, so that the thermodynamic constraints on H_2 formation are lessened thereby increasing its production. The concept can be applied to any stage of the process in which CO_2 acts as a limiting species (primary and secondary reformers or shift reactors) although it has been shown to be more suitable for the water gas shift (WGS) reaction [6]. The WGS is a common step in the synthesis of hydrogen from any carbon-containing source and is conventionally carried out in two stages: one at high temperature (573–773 K) with favourable kinetics but thermodynamically limited, and a low temperature step (483–543 K) with adequate equilibrium conversions but slow kinetics. In principle, only one reactor operating in the high temperature range is necessary in the sorption-enhanced water gas shift reaction (SEWGS) and therefore lower initial investments are expected. Additionally, the steam usage in the reactor is reduced and the purification of the hydrogen downstream is expected to be significantly simplified. The fact that CO_2 is removed in the process makes SEWGS an attractive pre-combustion CO_2 capture technology.

Besides sorption-enhancement, other process intensification strategies for syngas and hydrogen production have been discussed in the academic and the patent literature. Notable improvements have been achieved in the reformers by combining the endothermic steam methane reforming and the exothermic partial oxidation reactions into a more efficient autothermal process. Compact and efficient units have been designed in the form of shell and tube heat exchangers in which steam methane reforming takes place in the inside of the tubes and is followed by an autothermal converter. The hot effluent from the second stage provides the heat for the steam methane reforming [7]. Similar approaches have been suggested for the water gas shift stage. Since WGS is an exothermic reaction, the energy requirements can be reduced if the transformation is combined with endothermic processes such as steam-reforming of oxygenates [8, 9]. Recently Haldor Topsøe patented a process called methanol-to- shift (MTS) [10] which claims to increase the H_2 productivity and reduce the duty of the low temperature shift reactor by combining the WGS with the catalytic decomposition of methanol. The process is expected to be relevant in a context in which methanol is widely employed as energy carrier but it can be implemented using other oxygenates as feedstock. Furthermore, the methanol-to-shift process may be more competitive if it is combined with the in situ selective removal of the carbon dioxide formed (i.e. sorption-enhanced methanol- to-shift, SEMTS).

Sorption-enhanced shift reactions require the combination of a catalyst that is stable in the temperature range of interest (573–773 K), and a solid adsorbent with high CO_2 capacities and which must be easy to regenerate. While many catalyst systems appear as good candidates for the SEWGS processes, the selection of an adequate solid adsorbent remains a challenge. So far layered double hydroxides

(LDHs) are regarded as the most promising option since they show high selectivity towards CO_2 at temperatures between 473 and 723 K and good performance in the presence of water [11–14]. In addition compared to other potential chemisorbents such as calcium oxide and lithium zirconates, LDHs show faster adsorption-desorption kinetics and the energy required in their regeneration is significantly lower. However, despite these positive properties, the stability and CO_2 uptake of LDHs need to be further improved before they can be implemented in industry. An effective strategy to increase the CO_2 capacity of LDHs is to promote them with alkali ions but the effect of the promoters on the stability has not been studied sufficiently [12, 15, 16]. Alternatively, a novel approach suggests that the use of high surface area materials as a support for layered double hydroxides helps to increase the amount of CO_2 that is captured per unit of mass of LDH [17–19]. Regrettably, this effect has only been observed in studies with high support loadings so the capacities per total volume of adsorbent are rather low. Since in industry, the capacity per volume is the most important parameter as it dictates the size of the units, the supported LDHs reported so far have very limited applications.

1.2 Research Objectives

The present work is concerned with the production of hydrogen by the sorption-enhanced shift reactions that require the development of suitable catalysts and CO_2 solid adsorbents. In particular, the research concentrates on the study of layered double hydroxides and on strategies to improve their performance. For this purpose, the CO_2 adsorption characteristics of novel LDHs supported on multi-walled carbon nanotubes (MWCNTs) and graphene oxide (GO) are investigated. The materials were developed in collaboration with the Material Chemistry group in the Department of Chemistry at Imperial College London. Emphasis is placed on the assessment of the performance of the new LDH composites in terms of their CO_2 capacity and stability. In contrast to previous studies involving the use of supported LDHs, this research addresses the region of relatively low support loadings which is commercially more relevant. Additionally, the promotion of supported and unsupported LDHs with alkali ions is studied and their impact on the multicycle stability is evaluated. Detailed characterisation is carried out aiming to decouple the effects of supports and promoters on the performance of the adsorbents. This thesis also investigates for the first time the thermodynamic feasibility of the sorption-enhanced methanol-to-shift process. Preliminary screening of catalysts to be applied in SEWGS and SEMTS is conducted. The particular goals of the research are the following:

- Prepare composite adsorbents using supports that are compatible with the structure of LDH in terms of electric charge. MWCNTs were expected to provide a high surface area network for the dispersion of LDH crystallites while the accessibility of CO_2 to the adsorption sites is enhanced. On the other hand,

GO offers a 2D geometry which is more obviously compatible with the multilayer structure of LDH. In principle, a single nanosheet of GO is sufficient to stabilise one LDH platelet on each side suggesting a very high mass efficiency as support.

- Conduct a systematic characterisation of LDH supported on carbon nanostructures covering the regions of high and low support loadings. The objective is to understand the role of the support on the capacity and the stability of the adsorbents.
- Explore the promotion effect of sodium, potassium and caesium on unsupported and GO supported LDH and analyse the solids by different techniques to elucidate the nature of the promotion. The difference between adding the alkali after the synthesis and leaving them as residues from the coprecipitation is studied.
- Provide equilibrium and kinetic information that serve as a basis for the design of adsorption units. Studies are conducted using pure and GO supported LDHs under dry and wet conditions which are relevant for industrial applications. To this end a lab scale adsorption unit was designed and commissioned as part of this research. The preliminary modelling of the fixed-bed column aims to provide useful information for the prediction of the performance of larger units.
- Carry out a screening of viable catalytic systems to be used in the sorption-assisted processes of water gas shift and water gas shift in combination with methanol decomposition (methanol-to-shift). In the latter case, a thermodynamic study is done to evaluate the feasibility of a sorption-enhanced process.

1.3 Structure of the Thesis

This thesis is composed of nine chapters including this introduction. In Chap. 2 relevant background for the present work is reviewed covering some principles of gas-solid adsorption and the characteristics of promising groups of CO_2 adsorbents for carbon capture and storage. The CO_2 adsorption properties of layered double hydroxides (LDHs) and their derivatives produced upon thermal treatment (layered double oxides) are described in more detail. Additionally, the strategies reported so far to improve the adsorption performance of LDHs are discussed. The last part of the chapter focuses on hydrogen production highlighting the importance of process intensification by sorption-enhancement.

In Chap. 3 the standard procedures followed to test and characterise a range of LDH based adsorbents and catalysts are described. The use of a thermogravimetric analyser and the design and operation of a packed bed column used to carry out CO_2 adsorption measurements are explained in detail. In addition, the experimental setup used to conduct a preliminary screening of catalysts for the sorption-enhanced methanol-to-shift reaction is presented.

Chapters 4 and 5 are concerned with the study of layered double hydroxides supported on multi-walled carbon nanotubes and graphene oxide respectively. The chapters begin by describing the methodology used to synthesise LDHs and LDH/carbon hybrids. Subsequently, the structural and physical properties of the adsorbents are examined by a range of characterisation techniques. Finally the adsorption capacity and thermal stability of the materials are reported. Chapter 5 includes an additional section that deals with the description of CO_2 adsorption equilibrium data under dry conditions using Langmuir, Freundlich and Toth isotherms.

In Chap. 6 the effect of alkali metals on the CO_2 adsorption performance of unsupported and GO supported layered double hydroxides is investigated. A comparative study of the promoting effect of sodium, potassium and caesium on the adsorbents is presented. In addition, the influence of promoting the adsorbents by impregnation or by leaving residual alkali ions from the synthesis is discussed.

Chapter 7 addresses the CO_2 adsorption kinetics and equilibria of LDH and LDH/GO hybrids under dry and wet conditions using breakthrough curve responses obtained in a fixed-bed column. A comparative study between temperature-swing and isothermal N_2 purge experiments is presented. In addition, a mathematical model based on the linear driving force approximation is used to describe the dry experimental profiles.

Chapter 8 is divided in two main sections. The first part of the chapter presents a preliminary thermodynamic study to assess the viability of producing hydrogen by the sorption-enhanced methanol-to-shift (SEMTS). In the second section, the suitability of different solid catalysts for the SEWGS and SEMTS is evaluated by conducting the reverse water gas shift and methanol decomposition in temperature-programmed reaction experiments. Some catalysts were further studied in longer-term stability tests under water gas shift conditions.

Finally Chap. 9 presents the overall conclusions, contribution of the research and recommendations for future work.

References

1. Maitlis, P. M. (2013). What is Fischer–Tropsch? In *Greener fischer-tropsch processes for fuels and feedstocks* (pp 1–15). Weinheim: Wiley-VCH Verlag GmbH & Co. KGaA.
2. Subramani, V., Sharma, P., Zhang, L., & Liu, K. (2009). Catalytic steam reforming technology for the production of hydrogen and syngas. In *Hydrogen and syngas production and purification technologies* (pp 14–126). New York: Wiley.
3. Carvill, B. T., Hufton, J. R., Anand, M., & Sircar, S. (1996). Sorption-enhanced reaction process. *AIChE Journal, 42*(10), 2765–2772.
4. Harrison, D. P. (2008). Sorption-enhanced hydrogen production: A review. *Industrial and Engineering Chemistry Research, 47*(17), 6486–6501.
5. Stevens, R. W, Jr., Shamsi, A., Carpenter, S., & Siriwardane, R. (2010). Sorption-enhanced water gas shift reaction by sodium-promoted calcium oxides. *Fuel, 89*(6), 1280–1286.
6. Cobden, P. D., van Beurden, P., Reijers, H. T. J., Elzinga, G. D., Kluiters, S. C. A., Dijkstra, J. W., et al. (2007). Sorption-enhanced hydrogen production for pre-combustion CO_2 capture:

Thermodynamic analysis and experimental results. *International Journal of Greenhouse Gas Control, 1*(2), 170–179.
7. Zennaro, R., Ricci, M., Bua, L., Querci, C., Carnelli, L., & d'Arminio Monforte, A. (2013). Syngas: The Basis of Fischer–Tropsch. In *Greener fischer-tropsch processes for fuels and feedstocks* (pp 17–51). Weinheim: Wiley-VCH Verlag GmbH & Co. KGaA.
8. Comas, J., Laborde, M., & Amadeo, N. (2004). Thermodynamic analysis of hydrogen production from ethanol using CaO as a CO_2 sorbent. *Journal of Power Sources, 138*(1–2), 61–67.
9. Chen, H., Zhang, T., Dou, B., Dupont, V., Williams, P., Ghadiri, M., & Ding, Y. (2009). Thermodynamic analyses of adsorption-enhanced steam reforming of glycerol for hydrogen production. *International Journal of Hydrogen Energy, 34*(17), 7208–7222.
10. Nielsen, P. E. H., Hansen, J. B., & Schiødt, N. C. (2009). Process for the preparation of a hydrogen-rich stream. US 7,527,781 B2.
11. Ding, Y., & Alpay, E. (2000). Equilibria and kinetics of CO_2 adsorption on hydrotalcite adsorbent. *Chemical Engineering Science, 55*(17), 3461–3474.
12. Walspurger, S., Boels, L., Cobden, P. D., Elzinga, G. D., Haije, W. G., & van den Brink, R. W. (2008). The crucial role of the K^+–aluminium oxide interaction in K^+-promoted alumina- and hydrotalcite-based materials for CO_2 sorption at high temperatures. *ChemSusChem, 1*(7), 643–650.
13. van Selow, E. R., Cobden, P. D., Verbraeken, P. A., Hufton, J. R., & van den Brink, R. W. (2009). Carbon capture by sorption-enhanced water–Gas shift reaction process using hydrotalcite-based material. *Industrial and Engineering Chemistry Research, 48*(9), 4184–4193.
14. Wu, Y. J., Li, P., Yu, J. G., Cunha, A. F., & Rodrigues, A. E. (2013). K-promoted hydrotalcites for CO_2 capture in sorption enhanced reactions. *Chemical Engineering and Technology, 36*(4), 567–574.
15. Reijers, H. T. J., Valster-Schiermeier, S. E. A., Cobden, P. D., & van den Brink, R. W. (2005). Hydrotalcite as CO_2 sorbent for sorption-enhanced steam reforming of methane. *Industrial and Engineering Chemistry Research, 45*(8), 2522–2530.
16. Oliveira, E. L. G., Grande, C. A., & Rodrigues, A. E. (2008). CO_2 sorption on hydrotalcite and alkali-modified (K and Cs) hydrotalcites at high temperatures. *Separation and Purification Technology, 62*(1), 137–147.
17. Othman, M. R., Rasid, N. M., & Fernando, W. J. N. (2006). Mg–Al hydrotalcite coating on zeolites for improved carbon dioxide adsorption. *Chemical Engineering Science, 61*(5), 1555–1560.
18. Meis, N. N. A. H., Bitter, J. H., & de Jong, K. P. (2009). Support and size effects of activated hydrotalcites for precombustion CO_2 capture. *Industrial and Engineering Chemistry Research, 49*(3), 1229–1235.
19. Aschenbrenner, O., McGuire, P., Alsamaq, S., Wang, J., Supasitmongkol, S., Al-Duri, B., et al. (2011). Adsorption of carbon dioxide on hydrotalcite-like compounds of different compositions. *Chemical Engineering Research and Design, 89*(9), 1711–1721.

Chapter 2
Literature Review

2.1 Introduction

The present work involves the study of layered double hydroxides as CO_2 solid adsorbents with application in sorption-enhanced hydrogen production processes. In this chapter some relevant aspects are addressed including basic background on gas–solid adsorption, CO_2 adsorbents for carbon capture and storage, CO_2 adsorption characteristics of layered double hydroxides and the production of H_2 by sorption-enhancement.

Section 2.2 reviews some basic principles such as the definition of gas-solid adsorption, adsorption isotherms and mass and heat transfer phenomena. Relevant experimental methods to study adsorption (thermogravimetry and breakthrough curve analyses) are briefly explained. In Sect. 2.3 the characteristics of promising types of CO_2 adsorbents are mentioned including their equilibrium capacities and adsorption kinetics, the typical operating window, their performance in the presence of water and their regenerability. Section 2.4 highlights the characteristics and adsorption properties of layered double hydroxides and their derivatives. In addition, different strategies to enhance their performance are presented including the promotion with alkali metals and the use of high surface area materials as supports. In Sect. 2.5 some relevant properties of carbon nanostructures and their potential use as supports are reviewed. Subsequently, Sect. 2.6 describes the industrial production of hydrogen and discusses the relevance of process intensification by sorption-enhancement. Finally, Sect. 2.7 gives some concluding remarks.

2.2 Gas–Solid Adsorption

An increasing interest in gas adsorption processes in chemical and petrochemical industries has been observed since the second half of the last century fostered by numerous scientific and engineering developments such as the discovery of new

D. Iruretagoyena Ferrer, *Supported Layered Double Hydroxides as CO_2 Adsorbents for Sorption-enhanced H_2 Production*, Springer Theses,
DOI 10.1007/978-3-319-41276-4_2

porous materials (e.g. carbon molecular sieves and synthetic zeolites) and the invention of clever and novel processes (e.g. more efficient pressure-swing adsorption cycles) [1]. Nowadays, adsorption is of great importance representing not only a key tool for many existing industrial systems but also playing a crucial role in the development of new technologies. A deep understanding of the adsorption phenomena is very important and therefore many books have been written regarding this topic [1–8]. A brief description of some basic adsorption principles will be given in the following subsections.

2.2.1 *Basic Adsorption Principles*

The term adsorption is used to describe the phenomenon in which molecules of fluid phases, i.e. gases, vapours and liquids (*adsorptive*), enrich the surface of solids (or other liquid phases) forming an interface, that is considered a separate phase from the point of view of thermodynamics [9]. The set of adsorbed molecules (*admolecules*) on the surface of the solid material is called *adsorbate* whereas the solid phase, usually porous, is named *adsorbent* (Fig. 2.1). The phenomenon of desorption occurs when the adsorbed molecules return to the fluid phase. If the molecular flows to and from the surface are equal then adsorption equilibrium exists, otherwise either adsorption or desorption controls the process [8].

2.2.1.1 Heat of Adsorption

Adsorption phenomena usually exhibit an exothermic behaviour which can be easily explained by simple thermodynamic principles. An adsorbed molecule has less degrees of freedom on the surface of the solid than in the gas phase. As a

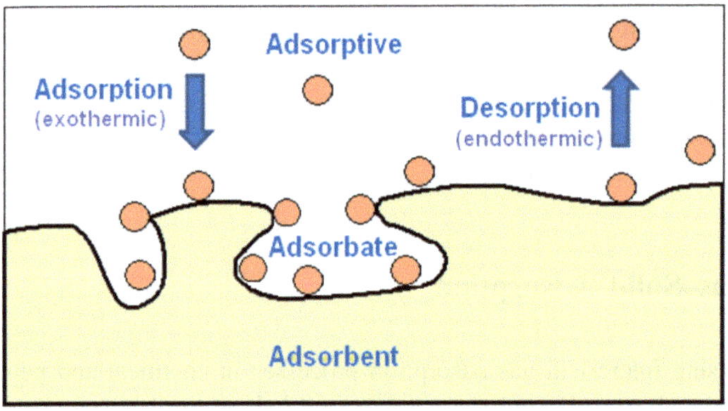

Fig. 2.1 Adsorption system consisting of sorptive gas (adsorptive), a sorbed phase or adsorbate (adsorbate) and a solid sorbent phase (adsorbent) (adapted from Keller [8])

consequence the entropy change caused by adsorption $\left(\Delta S_{ads} = S_{adsorbate} - S_{gas}\right)$ is necessarily negative. Spontaneous adsorption requires a negative free energy change $(\Delta G_{ads} = \Delta H_{ads} - T\Delta S_{ads})$, hence the enthalpy of adsorption (ΔH_{ads}) must be negative, i.e. the adsorption process is exothermic [3]. This implies that desorption is endothermic.

The heat of adsorption is defined as $Q_{ads} = -\Delta H_{ads}$ and therefore it is a positive quantity. Many definitions of heat of adsorption can be found in the literature [4, 10, 11]. Among them, the isosteric heat of adsorption is very useful for the design of adsorption units. It can be directly measured using calorimetric methods or estimated indirectly from experimental isotherms at different temperatures [10]. From measurements of pressure and temperature at constant coverage θ_i, the isosteric heat of adsorption can be calculated (using van't Hoff equation) as follows:

$$\left[\ln\left(\frac{p_1}{p_2}\right)\right] = \frac{-Q_{ads}}{R}\left(\frac{1}{T_1} - \frac{1}{T_2}\right) \tag{2.1}$$

2.2.1.2 Physical and Chemical Adsorption

A variety of phenomena can be expected to occur during adsorption due to the complex nature of the interactions of the adsorbate with the adsorbent or with co-adsorbed species. Depending on the strength or interaction energy by which admolecules are bound to the surface of the adsorbent, two main types of phenomena can be distinguished: physisorption and chemisorption. In the former, weak intermolecular forces are involved (e.g. van der Waals and/or dispersion forces due

Table 2.1 Basic properties of physisorption and chemisorption phenomena (Ruthven [3], Keller [8])

Property	Physisorption	Chemisorption
Selectivity of adsorptive gases	Non specific	Highly specific
Intensity of adsorption increased for adsorptive gas pressure and temperature	Favoured at high pressure and low temperature	Possible over a wide range of temperatures
Kinetics	Rapid	May be slow
Desorption	Reversible	Irreversible/reversible
Heat of adsorption	Low heat of adsorption (<2 or 3 times the latent heat of evaporation)	High heat of adsorption (>2 or >3 times higher the latent heat of evaporation)
Adsorbate-adsorbent interaction	Monolayer or multilayer, no dissociation of adsorbed species, no electron transfer although polarisation of adsorbate may occur	Monolayer only, may involve dissociation, electron transfer leading to bond formation between adsorbate and surface
Type of phenomenon	Non-activated	Activated/non-activated

to induced dipole-dipole interactions) whereas in the latter, strong bonds are created (e.g. covalent and metallic-type bonding) [11]. Although this distinction is conceptually useful, there are many intermediate cases and it is not always possible to categorise a particular system unequivocally [3]. The general features which differentiate physical and chemical adsorption are summarised in Table 2.1. Most of the industrial separation processes depend on physical adsorption rather than on chemisorption. This is due to the reversible character of physisorption, which allows the adsorbent to be more easily regenerated.

2.2.2 *Regeneration of Spent Adsorbents*

Separation by adsorption is an unsteady process in which regeneration of the adsorbent is needed for cyclic use. The main purpose of regeneration is to restore the adsorption capacity of the spent adsorbent and to recover any valuable component retained. A specific and finite capacity exists for each solid-adsorbate couple at specific temperature, pressure and gas composition. The saturated solids can be regenerated by conveniently changing at least one of these process parameters. Pressure-swing adsorption (PSA) and temperature-swing adsorption (TSA) are among the most used methods to regenerate solid adsorbents [4, 12].

2.2.2.1 Pressure-swing Adsorption

Pressure-swing adsorption processes (PSA) are commonly used for gas-solid systems involving physical adsorption where the regeneration of the adsorbent can be easily done by reducing the total pressure of the bed (i.e. the pressure of the system changes between high pressure during adsorption and low pressure during regeneration). PSA cycles are usually employed when the concentration of the components to be removed is significant (more than a few percent). Under those conditions, the adsorbent is saturated very fast and since the pressure of the system can be changed rapidly, the time between adsorption and regeneration is balanced allowing for a complete cycle in minutes or seconds. PSA technology is widely used in very large scales such as hydrogen recovery, air separation, CO_2 removal, purification of noble gases, CH_4 upgrading and n-iso paraffin separation [1, 5, 12].

2.2.2.2 Temperature-Swing Adsorption

In temperature-swing adsorption (TSA) the solid is primarily regenerated by heating which is usually provided by a purge gas at high temperature. TSA is particularly convenient in the separation of molecules that form strong bonds with the adsorbent. Therefore, this regeneration process has been frequently used to remove components present in relatively low concentrations such as organic solvents from air and moisture from natural gas. In this case, when high product purities are not achievable with PSA, TSA may be adequate. On the other hand, a challenge of this method is the long duration of the heating-cooling cycles which may adversely affect the cycle productivity. In practice, it is occasionally necessary to combine TSA and PSA cycles to achieve an effective regeneration of the solid [1, 4].

2.2.3 *Adsorption Isotherms*

The study of adsorption equilibrium is based on the understanding of adsorption isotherms, which establish the relation between the amount adsorbed and the partial pressure of a species at constant temperature. Isotherms are determined experimentally being usually depicted as a plot of fractional coverage of adsorbate (θi) or the amount adsorbed (qi) versus the relative pressure (p_i/P_o) of the gas. A wide variety of mathematical models have been formulated to interpret different types of isotherms including the equations of Langmuir, Freundlich, Langmuir-Freundlich (Sips), Toth, Temkin, Dubinin and Brunauer-Emmet-Teller (BET). In the following paragraphs the models relevant to the present study are briefly described.

2.2.3.1 Langmuir Isotherm

In 1918, Langmuir proposed the first theoretical model for monolayer adsorption to describe mainly chemisorption phenomena. The basic assumptions of the Langmuir model are: (1) the surface of the adsorbent is homogeneous (i.e. all sites are energetically equivalent), (2) adsorption on the surface is localised (i.e. the molecules are adsorbed on a fixed number of well-defined localised sites), (3) each site can accommodate only one adsorbate molecule and (4) there is no interaction between molecules adsorbed on neighbouring sites [3, 6].

The original model that Langmuir developed was based on the kinetic theory of gases. However, the isotherm can also be derived by other methods including statistical mechanics, thermodynamics and chemical reaction equilibrium [13]. The last approach is relatively simple and useful, and it is developed as follows:

The nondissociative chemisorption of a gas (A) on a site (S) can be represented as:

$$A_{(g)} + S \quad \mathrm{k}_{ads} \leftrightarrow \mathrm{k}_{des} \quad A \cdot S$$

$$\text{Rate of adsorption} \quad k_{ads} p_A (1 - \theta_A) \tag{2.2}$$

$$\text{Rate of desorption} \quad k_{des} \theta_A \tag{2.3}$$

$$\theta_A = \frac{q_A}{m} \tag{2.4}$$

where k_{ads} and k_{des} are the adsorption and desorption rate constants respectively, p_A is the partial pressure of A, q_A is the amount of A adsorbed at a fixed set of temperature and p_A, m is the monolayer capacity (i.e. the amount of A required to completely cover the surface of the adsorbent, which should be independent of temperature) and θ_A is the fractional coverage (dimensionless). At equilibrium the rates of adsorption and desorption are equal:

$$k_{ads}p_A(1-\theta_A) = k_{des}\theta_A \tag{2.5}$$

Equation (2.5) can be rearranged to obtain the common expression of the Langmuir isotherm:

$$\theta_A = q_A/m = bp_A/(1+bp_A) \tag{2.6}$$

In Eq. 2.6, $b = k_{ads}/k_{des}$ is the adsorption equilibrium constant. When the partial pressure of A approaches infinity, the fractional coverage tends to 1, while at low adsorbate concentrations ($bp_A \ll 1$), the Langmuir expression reduces to the Henry's law isotherm:

$$\theta_A = q_A/m = bp_A \tag{2.7}$$

The dependence of the adsorption equilibrium constant with temperature follows van't Hoff equation:

$$-RT\ln(b) = \Delta G_{ads} = \Delta H_{ads} - T\Delta S_{ads} \tag{2.8}$$

Therefore b can be expressed as:

$$b = \exp\left(\frac{\Delta S_{ads}}{R}\right)\exp\left(\frac{-\Delta H_{ads}}{RT}\right) = b_o\exp\left(\frac{-\Delta H_{ads}}{RT}\right) = b_o\exp\left(\frac{Q_{ads}}{RT}\right) \tag{2.9}$$

where ΔH_{ads} is the enthalpy of adsorption (negative) and Q_{ads} is the heat of adsorption (positive), which for the Langmuir's model are independent of the coverage. The parameter b, also called affinity constant, is a measure of how strong an adsorbate is attracted onto a surface and decreases with increasing temperature since more energy is provided to desorb the adsorbed molecules.

2.2.3.2 Freundlich Isotherm

In many cases, the assumptions involved in the derivation of the Langmuir model are not adequate to describe real surfaces. Therefore various empirical equations have been proposed to deal with systems that deviate from an ideal behaviour. Among the earliest efforts is the isotherm suggested by Freundlich in 1926, which is expressed as:

$$q_A = kp_A^{1/n} \quad \text{or} \quad \theta_A = k' p_A^{1/n} \tag{2.10}$$

where k and n are generally temperature dependent, decreasing with increasing temperature [14].[1] The parameter n is usually greater than unity and the larger this value is, the more nonlinear the adsorption isotherm becomes [6]. The Freundlich isotherm is expected to be valid only for moderate coverages since the equation does not have a proper Henry's law behaviour at low pressure and it does not have a finite limit when the pressure is very high.

Although this model was proposed originally as an empirical isotherm, it has been justified by thermodynamic and statistical approaches [13, 15]. The theoretical derivation of the Freundlich isotherm considers that the surface of the adsorbent is heterogeneous and can be divided into small domains that contain sites with the same adsorption energy. Each domain is independent from each other (i.e. there is no interaction between domains) and on each domain the adsorbate is adsorbed on one site exclusively. Therefore, the adsorption on a single site within each domain, i, can be described by Langmuir:

$$\theta_i = b_i p_A / (1 + b_i p_A) \tag{2.11}$$

If the bi values (defined in Eq. 2.9) for each domain are assumed to be distributed continuously, the overall coverage can be written as:

$$\theta = \int \alpha_i \theta_i d_i \tag{2.12}$$

where α_i is the fraction of type i sites and $\alpha_i d_i$ is the frequency of occurrence of θ_i between i and $i + \Delta i$. If it is assumed that the domains i can be represented by the heat of adsorption, Q_{ads}, and that α_i depends exponentially on Q_{ads} then:

$$\alpha_{Q_{ads}} = \alpha_0 \exp(-Q_{ads}/Q_{ads0}) \tag{2.13}$$

where Q_{ads0} and α_0 are constants. Substituting Eqs. (2.13) and (2.11) into Eq. (2.12) and integrating results in the following expression for the coverage:

$$\alpha_{Q_{ads}} = \alpha_0 \exp(-Q_{ads}/Q_{ads0}) \tag{2.14}$$

The fact that $\alpha_{Q_{ads}}$ represents a fractional distribution of adsorption sites requires that $\alpha_0 Q_{ads0} = 1$ [13]. Consequently, Eq. (2.14) can be expressed as $\theta = (b_0 p_A)^{RT/Q_{adso}}$ which coincides with the empirical Freundlich isotherm (Eq. 2.10). The detailed derivation is given in Appendix A.1.

[1] The dependence of k and n on temperature is complex so they should not be extrapolated outside their range of validity. Usually $\ln(k)$ and $1/n$ depend linearly on temperature (Huang and Cho [14]).

2.2.3.3 Toth Isotherm

To broaden the pressure range where the Freundlich model can be applied, other empirical isotherms have been proposed such as the Sips (Langmuir-Freundlich) and Toth equations. The latter isotherm is popular since it is valid at high partial pressures and possesses the correct Henry's law type behaviour at low pressure. The Toth Equation (1971) is a three parameter model and has the following form:

$$q_A = mbp_A/\left(1 + (bp_A)^t\right)^{1/t} \tag{2.15}$$

where b and t are specific for each adsorbate-adsorbent pair and depend on temperature [6]. The parameter t is usually less than unity and characterises the heterogeneity of the system. Deviations of t from unity correspond to heterogeneous systems and may stem from the nature of the solid or the adsorbate, or from a combination of both. When $t = 1$, the Toth isotherm reduces to the Langmuir model which represents a homogenous system. Similarly to the Freundlich isotherm, the Toth equation can also be theoretically justified [6]. The Toth model is used to describe experimental adsorption isotherms of hydrotalcite based adsorbents in Chap. 5.

2.2.4 *Mass Transfer Phenomena*

Before a molecule of gas adsorbs on a solid particle, it must be transported from the well-mixed gas phase to the exterior surface of the adsorbent and, if the material is porous, further into the pore system. External mass transfer resistances exist when a stagnant film of gas is formed around the adsorbent particles creating a concentration gradient between the homogeneous bulk phase and the exterior surface of the solid. On the other hand, if the diffusion inside the adsorbent particles is limited, differences in concentration of the adsorptive gas will develop between the external surface and the interior of the pores.

In the development of mathematical models to describe adsorption in packed columns, it is of great interest to estimate the relative contribution of internal and external phenomena to the overall mass transport. This can be achieved using the *Biot* number, which is a dimensionless quantity that compares the internal and external mass transfer resistances as follows:

$$Bi_m = \frac{\text{internal resistance}}{\text{external resistance}} = \frac{k_m L_C}{D} \tag{2.16}$$

where Bi_m is the mass transfer *Biot* number, k_m is the mass transfer coefficient for the external stagnant film, L_C is a characteristic dimension and D is a term that considers the overall diffusivity of the adsorptive through the pore system [6]. In general, adsorbers operate under conditions where the *Biot* number is large

($Bi_m \gg 1$) and therefore external transfer limitations are usually neglected in the mathematical descriptions [1, 6].

Regarding the resistance to mass transport through the pores, many different diffusion mechanisms may occur simultaneously. Inside very wide pores, the transport is governed by the molecular diffusivity of the adsorptive in the gas mixture. As the pore diameter approaches the mean free path of the gas molecules, interactions with the pore walls become important and the process is controlled by Knudsen diffusion. Additionally, once the molecules are adsorbed, surface diffusion takes place when the adsorbate overcomes the energy barrier to desorb from one adsorption site and hops to an adjacent site. As a consequence, the description of the intraphase transfer requires an expression (D_e, effective diffusivity) that accounts for the different mechanisms that take place during the diffusive transport.

To model adsorption in a packed column, the mass balance within the spherical pellet and for the bulk flow of the bed need to be considered using the following equations [1].

Spherical pellet mass balance:

$$D_e\left(\frac{\partial^2 c_{i_{pore}}}{\partial r^2} + \frac{2}{r}\frac{\partial c_{i_{pore}}}{\partial r}\right) = \frac{\partial q_i}{\partial t} \tag{2.17}$$

Bulk flow mass balance:

$$-D_z\frac{\partial^2 c_{i_{bulk}}}{\partial z^2} + \frac{\partial v c_{i_{bulk}}}{\partial z} + \frac{\partial c_{i_{bulk}}}{\partial t} + \frac{1-\varepsilon}{\varepsilon}k_m a\left(c_{i_{bulk}} - c_{i_{surf}}\right) = 0 \tag{2.18}$$

Continuity at the pellet surface:

$$D_e\left(\frac{\partial c_{i_{pore}}}{\partial r}\right)_{r=R_p} = k_m\left(c_{i_{bulk}} - c_{i_{surf}}\right) \tag{2.19}$$

where D_e and D_z are the effective and axial diffusivities respectively (m^2 s^{-1}); $c_{i_{pore}}$, $c_{i_{bulk}}$ and $c_{i_{surf}}$ are the adsorptive gas concentrations of species i in the pores, the interpellet region and the external surface of the particle respectively (mol$_{gas}$ m^{-3}); v is the interstitial velocity (m s^{-1}), ε is the interpellet void fraction; k_m is the average mass transfer film coefficient (m s^{-1}) and a is the exterior surface area of pellets per volume of bed (m^2 m^{-3}).[2] However, the solution of these mass balance equations is complicated and demands a tremendous amount of computational time. For this reason, simplifying assumptions are usually made to eliminate the mass balance equation within the adsorbent particles and substitute it by an expression that relates the overall uptake rate to the bulk flow concentrations. Among the most

[2]In Eq. (2.17) the adsorbate accumulation in the gas phase, $\varepsilon_p \partial c_{i_{pore}}/\partial t$ (where ε_p is the pellet porosity), has been neglected otherwise it appears on the right-hand side.

used approximations for mass transfer rate is the linear diving force model, which is expressed as:

$$\frac{\partial q_i}{\partial t} = k_{LDF}\left(q_i^* - q_i\right) \tag{2.20}$$

where q_i is the overall amount of species i adsorbed averaged over the particle volume ($\text{mol}_{\text{gas}}\ \text{m}_{\text{ads}}^{-3}$), q_i^* is the amount of i adsorbed in equilibrium with the bulk phase concentration, $c_{i_{bulk}}$, and k_{LDF} (s^{-1}) is a parameter that lumps the different mass transport phenomena that take place within the adsorbent particles. Equation (2.20) replaces Eq. (2.19) and is substituted in the last term on the right-hand side of Eq. (2.18). The pellet mass balance (Eq. 2.17) is thereby eliminated from the model and the resulting material balance Eq. (2.21) can then be solved with the appropriate initial and boundary conditions.

$$-D_z \frac{\partial^2 c_{i_{bulk}}}{\partial z^2} + \frac{\partial v c_{i_{bulk}}}{\partial z} + \frac{\partial c_{i_{bulk}}}{\partial t} + \frac{1-\varepsilon}{\varepsilon} k_{LDF}\left(q_i^* - q_i\right) = 0 \tag{2.21}$$

The linear driving force approximation has been widely used to model adsorbers as well as gas separation breakthrough curves, desorption behaviour and cyclic separation processes. In this thesis the LDF model is used to describe the adsorption kinetics of CO_2-hydrotalcite systems in Chap. 7.

2.2.5 Heat Transfer Phenomena

The effects of heat generation and heat transfer in an adsorptive system need to be taken into account since they affect the adsorption equilibrium and in some cases the adsorption rate. The heat released during transient adsorption gives rise to non-isothermal profiles inside and outside the adsorbent particles. This heat is transferred by conduction between the solid particles and by convection from the adsorbent to the fluid medium; the radiation term is usually neglected. Similar to mass transfer phenomena, there are heat transfer resistances which occur in the external gas film and/or within the particle itself. Depending on which of these limitations are relevant, heat transfer expressions analogous to those written above for mass transfer need to be used [1, 3, 6]. The equations governing heat and mass transport phenomena need to be solved simultaneously to fully describe the adsorption process.

2.2.6 Experimental Methods to Study Adsorption

The continuous progress of adsorption knowledge has led to the development of many experimental methods to study gas–solid adsorption. A consensus to organise these techniques has not been established yet and therefore many different classifications can be found in the literature [1, 2, 8]. In general they can be grouped either as static or dynamic methods. In the static techniques the gas is brought into contact with the adsorbent in successive doses, either directly or through a capillary. Most of the existing static apparatus (volumetric and gravimetric) are planned to work at low pressure although they can also be designed to work at high pressure [16, 17]. A difficulty associated with these type of methods is that, in order to obtain accurate data, the time required for the equilibration of the system is usually in the order of days. On the other hand, in the dynamic methods the gas flows over the adsorbent for the duration of the experiment. The main advantages of using these transient techniques are that the apparatus are simple and they can be designed to operate over a wide range of pressures and temperatures. In addition, contrary to static methods that usually provide equilibrium data exclusively, information related to the adsorption-desorption kinetics can also be obtained. Among the different dynamic techniques, thermogravimetric analysis (TGA) and adsorption in packed bed columns are relevant for the present research.

2.2.6.1 Thermogravimetric Analysis

The TGA technique is one of the most practical ways to obtain adsorption data. The amount of gas adsorbed per weight of material (adsorption capacity) is determined by a micro-balance, which measures the weight gain as gas molecules are adsorbed on the solid material (at a particular temperature and partial pressure). Since the changes in weight can be very small, they have to be contrasted always to the sensitivity limits of the instrument so that the reliability of the data can be determined. The method also enables to measure adsorption-desorption kinetics [18, 19], although the parameters obtained are considered to be only estimates of those that correspond to packed bed configurations. This is because usually in the TGA apparatus the gas is flowed past and no through an adsorbent bed, which may slow down the transfer of the adsorptive gas to the active adsorption sites [20].

2.2.6.2 Packed Bed Column

Fixed-bed systems have been widely used to measure gas adsorption on solids due to their relatively easy operation and construction. One of the most common methods to study the dynamics of adsorption columns is to monitor the response of an initially adsorbate free bed to an imposed stimulus in the feed composition of the adsorptive gas. The concentration of the effluent is typically followed by means of a mass spectrometer (MS), infrared spectrometer (IRS), thermal conductivity detector (TCD) or gas chromatograph (GC). The first three detectors allow data to be measured in real time (order of seconds), whereas the frequency of measurements with a GC is usually limited by the time required for the gas sample to pass through the GC column. If the stimulus applied corresponds to a step change, the response is often

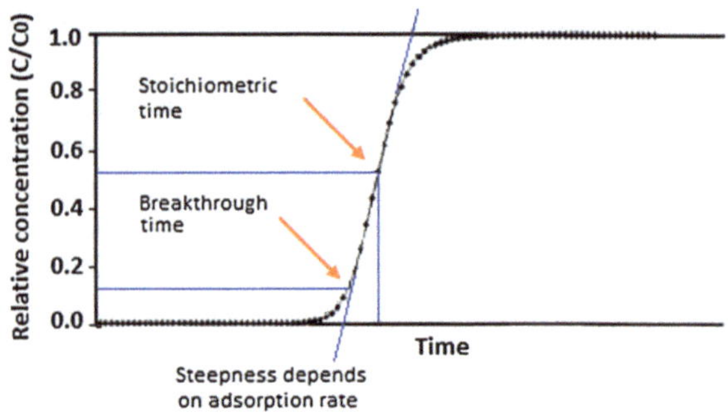

Fig. 2.2 Characteristics of a typical adsorption breakthrough curve (adapted from Wood et al. [21])

referred as *breakthrough curve*. This transient profile reflects the progress of the adsorbent saturation and mass transfer zone (MTZ)[3] through the column and therefore valuable equilibrium and kinetic information can be extracted from it [3, 4].

A typical breakthrough curve is depicted in Fig. 2.2 and has four distinctive features: (1) the geometric midpoint (stoichiometric time and the corresponding relative concentration), (2) the steepness, (3) the shape and (4) the breakthrough point [21]. The geometric midpoint depends on the superficial velocity of the flowing gas mixture, on the characteristics of the packed bed (size and voidage) and on the adsorption capacity of the material used. The steepness of the elution front is related to the rate at which the adsorptive is removed from the carrier gas. If the overall mass-transfer rate coefficient is large and the approach to equilibrium occurs very rapidly, the breakthrough curve will be very steep while slower uptake rates will give more gradual responses. The shape of the breakthrough curve is determined mainly by the characteristics of the adsorbent. Adsorption on homogeneous surfaces and with constant mass-transfer coefficients along the bed gives symmetrical responses. On the contrary, the asymmetric shape of the breakthrough curves is attributed to the heterogeneity of the adsorbents since the forefront of the adsorption wave will occupy the most active and accessible sites leaving the less active ones for subsequently arriving molecules [21]. The breakthrough point is defined when a fixed concentration of the adsorptive gas emerges in the exit of the column, and the effluent purity no longer meets the specifications of the process.

The geometric midpoint of the breakthrough curve is described by the stoichiometric time. For initially clean adsorbents, the stoichiometric time corresponds to the time at which the bed would be completely saturated if all the adsorptive molecules arrived to all sites simultaneously and the adsorption was instantaneous. It can be calculated by plotting the normalised approach of the effluent to the inlet concentration and integrating over time. That is:

[3]The term mass transfer zone (MTZ) refers to the section of the column where active adsorption takes place and it is commonly used to classify adsorption systems and to select appropriate models to describe their behaviour [3].

$$t_{st} = \int_{t_0}^{t_\infty} (1 - c_i/c_{i_0})dt \tag{2.22}$$

where t_{st} is the stoichiometric time and c_i and c_{i_0} are the concentrations of adsorptive at the outlet and the inlet of the packed bed respectively. The lower integration limit, t_0, corresponds to the instant when the step change is done while the upper limit, t_∞, is the time necessary so that the feed and the effluent concentrations are equal. The stoichiometric time is often used in expressions to calculate the total adsorption capacity of packed columns [22, 23].

2.2.7 Solid Adsorbent

The selection of a proper porous adsorbent is crucial in all adsorptive processes since good performances in both kinetics and equilibria are required. The kinetic behaviour of the solid is related to the intraparticular diffusion rates of the different adsorbates, whereas the equilibrium is associated with the adsorption capacity of the material and its selectivity towards a particular molecule (depending on the equilibrium affinity). A solid with high capacity but slow kinetics is not suitable since long residence times are needed, i.e. much time is required for the gas to get into the interior of the particle. On the other hand, fast kinetics but small adsorptive capacity is not adequate either as large amounts of material are required for acceptable adsorption yields. Therefore a good adsorbent must have high adsorption capacity and also good kinetics; otherwise low adsorption throughputs are obtained [6].

In order to fulfil the adsorption capacity requirement, the adsorbent must have an adequate micropore volume or specific surface area (generally well in excess of 100 m^2 g^{-1}). In addition relatively large pore networks favour kinetics since the diffusing molecules readily reach the micropores in the interior of the particle. This implies that the solid must have a combination of different pore sizes: micropores ($d_{pore} < 2$ nm), mesopores ($2 < d_{pore} < 50$ nm) and macropores ($d_{pore} > 50$ nm) [6, 24].

Besides a suitable porosity, other criteria are essential to select an effective adsorbent. The solid needs to exhibit a reversible adsorption behaviour being stable after repeated adsorption-desorption cycles. Additionally an adsorbent with good mechanical strength is required to work under high pressure. In some systems the hydrophobicity of the materials is very important since the adsorptive gas may contain moisture. Popular commercial solids which usually satisfy these conditions are: alumina, silica gel, activated carbon and zeolites.

2.3 Carbon Dioxide Capture by Solid Adsorbents

As mentioned in Chap. 1, over the past century there has been a dramatic increase in the concentration of CO_2 in the atmosphere, which in turn has made a significant contribution to global warming. The main source of CO_2 associated with human activities is fossil fuel combustion used for power and heat generation and for transportation, which in total accounts for almost 70 % of the global emissions of CO_2 [25]. Unfortunately, an immediate CO_2-emission halt is not possible since energy demands are expected to increase significantly during the next decades and the so-called green energies are not sufficiently developed to replace fossil fuels on a large scale [26]. As a consequence major technological efforts continue to be devoted to finding suitable processes for carbon dioxide capture and storage (CCS). Gas-solid adsorption is one of the most promising strategies for post-combustion and pre-combustion CO_2 capture applications.[4] Unlike liquid sorbents, solid adsorbents can be used over a wide temperature range. In addition, CO_2 solid adsorbents that operate at mid- and high- temperatures can be used to enhance the production of H_2 by the so-called sorption-assisted catalytic reactions, which are also regarded as attractive pre-combustion technologies (details are given in Sect. 2.5.1).

An ideal CO_2 adsorbent must fulfil all the conditions mentioned in Sect. 2.2.7. Therefore it must have high selectivity and adsorption capacity for carbon dioxide, fast adsorption-desorption kinetics, adequate multicycle stability and good performance in the presence of water and/or other competing species. In response to these demanding requirements, a range of potential CO_2 adsorbents has been investigated in recent years. Since each has its weaknesses and strengths, the selection of an effective CO_2 adsorbent must be based on giving priority to some criteria so that the constraints of a specific adsorptive process can be met.

All the promising adsorbents that have been identified so far can generally be grouped according to the range of temperature in which they exhibit best performance. Some physisorbents such as zeolites [27–29] and activated carbons [30–32] have been reported to be competitive at temperatures below $\sim$393 K. Other materials are also effective at low temperature such as organic-inorganic hybrids (e.g. amines impregnated in silica or covalently bound to silica) [33–35] and an emerging class of adsorbents called metal-organic frameworks [36–38].

Although some authors have suggested the use of zeolites and activated carbon at temperatures near 523 K [39, 40], some chemisorbents clearly perform better under those conditions. Among these chemisorbents, some undergo surface chemical reactions with CO_2 while others experience bulk chemical reactions. The

[4]In post-combustion CO_2 capture technologies CO_2 is separated from a flue gas after the combustion of a fossil fuel (e.g. the removal of CO_2 from coal-fired power plants). Pre-combustion CO_2 capture refers to processes where CO_2 is removed before the combustion of the fossil fuel (e.g. the fuel is converted to syngas and subsequently to a H_2-rich stream from which CO_2 is captured).

former group exhibits good adsorption parameters at temperatures between 473 and 773 K (e.g. porous magnesium oxide [41, 42], promoted aluminas [43, 44] and layered double hydroxides [18, 23, 45, 46]), whereas the latter group of adsorbents is suitable in a temperature range from 673 to 973 K (e.g. lithium zirconates [47–49] and calcium oxides [50–52]). Table 2.2 provides a summary of adsorption characteristics of various groups of promising CO_2 adsorbents. In the next section a detailed description of layered double hydroxides (LDHs) is given since these materials are relevant for the present work. LDHs exhibit an adequate multicycle stability and their performance is enhanced in the presence of water. In addition, compared to lithium zirconates and calcium oxides, LDHs show faster adsorption-desorption kinetics and the energy required in their regeneration is significantly lower. Although LDHs show relatively low adsorption capacities, their positive CO_2 adsorption properties and their operating window make them very attractive for some flue gas CO_2 recovery processes and also for sorption-assisted applications such as the sorption-enhanced water gas shift reaction (SE-WGS) and the sorption-enhanced steam methane reforming reaction (SE-SMR) [53–56].

2.4 Layered Double Hydroxides

Layered double hydroxides (LDHs), also known as hydrotalcites (HTs) or mixed-metal layered hydroxides, belong to a large class of anionic clay minerals. Their two-dimensional structure is composed of positively charged brucite-like ($Mg(OH)_2$) layers in which divalent cations are partially substituted by trivalent cations. These positive ions are located at the centre of octahedral sites of hydroxide sheets, whose vertexes contain –OH anions. Each hydroxyl is shared by three octahedral cations and points towards the interlayer regions. The excess of charge is compensated by negative anions and water molecules located in the interlayer space. This region is highly disordered and its water content depends on the temperature, the water vapour pressure and the nature of the anions present [57]. The neutral structure of the LDH (Fig. 2.3) can be represented by the general formula:

$$\left(M_{1-x}^{2+}M_x^{3+}(\mathrm{OH})_2\right)^{x+}\left(A_{x/m}^{m-}\cdot n\mathrm{H_2O}\right)^{x-}$$

where x is the partial substitution of M^{2+} and M^{3+} and is usually between 0.17 and 0.33 (i.e. M^{2+}/M^{3+} rations between 2 and 5) [18, 57].

In the natural mineral LDH, the divalent metal is Mg^{2+}, the trivalent cation is Al^{3+} and the compensating anion is $CO_3{}^{2-}$ with the formula $Mg_6Al_2(OH)_{16}CO_3 \cdot 4H_2O$. Although these are the most common ions used in the preparation of synthetic LDHs, a large variety of LDHs have been obtained by using a wide range of metals (e.g. M^{2+}:Ni^{2+}, Zn^{2+}, Cu^{2+} and Mn^{2+}; M^{3+}:Fe^{3+}, Cr^{3+}, Co^{3+} and Ga^{3+}; A^{m-}:$SO_4{}^{2-}$, $NO_3{}^-$, Cl^- and OH^-) [18, 57, 58].

Table 2.2 Summary of the adsorption characteristics of promising CO_2 adsorbents

CO_2 adsorbent	Type of adsorption	Adsorption temperature (K)	Adsorption capacity (mol kg^{-1})	Adsorption-desorption kinetics
(a)				
Zeolites	Mainly physisorption	[298–393]	0.1 < a. c. < 5.5	Fast kinetics described by diffusion models. Equilibrium is achieved in minutes
Activated carbons	Mainly physisorption	[278–348]	0.1 < a. c. < 3.5	Fast kinetics described by diffusion models. Equilibrium is achieved in minutes
Amines impregnated in silica	Mainly chemisorption	[298- 343]	2 < a.c. < 5	Slow kinetics. Equilibrium usually achieved in several hours
Amines covalently bound to silica and organic supports (e.g. carbons, polymers, resins)	Mainly chemisorption	[298–343]	0.1 < a.c < 4 average ~ 1.2	Generally fast initial rate followed by slow rate controlled by diffusion. Equilibrium reached in few hours
Metal-organic frameworks (MOFs)	Physisorption	Room temp. [~298]	max a.c ~3	–
Layered double hydroxides (LDHs)	Physisorption/Chemisorption	[473–773]	0.1 < a.c < 1	Good kinetics described by diffusion models. Equilibrium usually achieved in few minutes/hours
Lithium zirconates	Chemisorption	[673–873]	0.5 < a.c < 6	Slow kinetics. Equilibrium usually achieved in several hours
Calcium oxide	Chemisorption	[723–973]	2 < a. c. < 11.5	Fast initial rate controlled by reaction, followed by slow rate controlled by diffusion. Equilibrium usually achieved in several hours

CO_2 adsorbent	Regeneration	Effect of steam	Structural modifications/promotion
(b)			
Zeolites	Easy to regenerate by PSA and TSA [373–573 K]	Usually unfavourable	Alkali and alkaline earth ionic exchange

(continued)

Table 2.2 (continued)

CO_2 adsorbent	Regeneration	Effect of steam	Structural modifications/promotion
(b)			
Activated carbons	Easy to regenerate by PSA [298 K]	Usually unfavourable	Incorporation of alternative functional groups on the carbon surface (such as amines)
Amines impregnated in silica	Moderate regenerability, usually regenerated by TSA ($\sim$343 K) and inert gas purge	Not clear. Generally unfavourable	Different amine loading and method to prepare the support
Amines covalently bound to silica and organic supports (e.g. carbons, polymers, resins)	Good regenerability by TSA ($\sim$373 K) and inert gas purge	Favourable	Different amine content and method to prepare the oxide or organic support
Metal-organic frameworks (MOFs)	–	–	Incorporation of amine functionalities giving place to chemisorption sites
Layered double hydroxides (LDHs)	Good regenerability by PSA and TSA [$\sim$673 K]	Favourable for adsorption-desorption	Change in framework anions and cations, promotion with alkali metals (e.g. K_2CO_3) and addition of supports
Lithium zirconates	Good regenerability at $\sim$1073 K	Not clear	Morphology and/or crystalline structure modification, addition of alkali components (e.g. Na_2ZrO_3, K_2CO_3)
Calcium oxide	Rapid degradation during multicycle adsorptive tests. Regeneration by calcination at $\sim$1173 K	Suitable for adsorbent regeneration	Use of rigid supports (e.g. Al_2O_3) and/or additives (e.g. alkali metals), preparation of nanoparticles, use of precursors different from natural $CaCO_3$ (e.g. $Ca(OH)_2$)

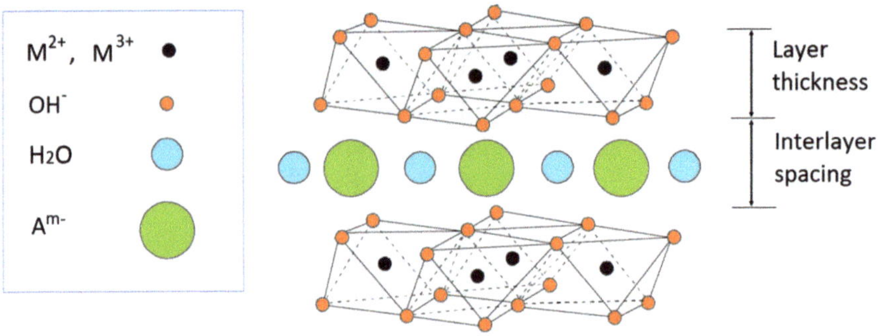

Fig. 2.3 Structure of layered double hydroxides (adapted from Hutson et al. [18])

Layered double hydroxides are typically synthesised by co-precipitation of aqueous solutions of M^{2+} and M^{3+} metal salts[5] in the presence of the desired interlayer anion. During the preparation of the LDHs, the pH needs to be maintained at higher (or equal) values than that at which the most soluble hydroxide precipitates [59]. Generally a pH range of 8–10 can be used to synthesise most LDHs [60]. The co-precipitation method can be performed under low or high supersaturation conditions. In the former case the two metal salts are added simultaneously to an alkaline solution to maintain the pH constant. On the other hand, in the high supersaturation method the M^{2+}/M^{3+} solution is added to a solution that contains the alkali and the anion precursors. In general, the precipitates obtained by high supersaturation are less crystalline than those produced by low supersaturation since the rate of nucleation tends to be higher than the rate of crystallite growth [59]. For both methods, co-precipitation is followed by a thermal treatment to enhance the yield and the crystallinity of the LDH. Besides co-precipitation, other methods have been used to synthesise LDHs such as the urea and sol-gel techniques [59–62].

Besides the potential application of LDHs as CO_2 adsorbents, their basic properties make them attractive for many catalytic reactions such as self-condensation, cross-aldol condensation of aldehydes and ketones, Michael addition, transesterification and alkylation. Additionally, they have a good anion-exchange capacity and therefore they are used as ion-exchangers and sensors. As mentioned above, LDHs can be prepared with many reducible divalent (e.g. Ni, Cu, Co) and trivalent cations (e.g. Fe, Cr) in the structure which give them suitable redox properties. The intimate contact between the metallic components of the LDHs that is achieved after calcination allows these structures to be used as precursors for the preparation of mixed oxides active for oxidation, hydrogenation and dehydrogenation reactions [63].

---

[5]Metal nitrate and chloride salts are commonly used due to the low selectivity of LDHs towards these anions [59].

2.4.1 *Effect of Thermal Treatment on the Structure of Layered Double Hydroxides*

Due to their relatively low basicity, as-synthesised LDHs are often inactive for CO_2 adsorption at high temperature and in many catalytic reactions. The poor basic character of fresh LDHs is mainly caused by the presence of adsorbed water which hinders the access of the adsorptive/reactive gas to the basic sites located on the surface and on the interlayer region [63]. Therefore hydrotalcites often need to be activated by a thermal treatment before being used. Typically LDHs are calcined in nitrogen or air between 673 and 823 K for at least 4 h [61, 64–66]. During this treatment, Mg–Al LDHs are transformed into layered double oxides (LDOs), which are well-dispersed Mg(Al)O mixed oxides with a high surface area that stems from the formation of significant porosity [67]. Although XRD measurements have shown that these derivatives are amorphous [61], SEM and TEM images have revealed that they usually maintain the morphology of the original precursor [68]. The use of excessive calcination temperatures leads to the appearance of spinel phases ($MgAl_2O_4$), which commonly should be avoided in order to preserve the platelet morphology of LDHs and an adequate basicity [63]. Costa et al. [65] found that a calcination temperature of 673 K produces LDOs with an optimum balance between basic sites and surface area which maximises their CO_2 adsorption capacity at temperatures above 473 K. It is worth mentioning that when LDOs are rehydrated, the double layers are partially reconstructed to produce meixnerite ($Mg_6Al_2(OH)_{18}{\cdot}4H_2O$), Fig. 2.4 [69]. This structure exhibits mild Brønsted basicity and is an efficient catalyst in many organic reactions such as self- and cross-aldol condensation of aldehydes and ketones [70–73].

Numerous studies have investigated the structural changes that LDHs experience with temperature [65, 67, 74–76]. Hibino et al. [74] published a detailed study investigating the decarbonation behaviour during heat treatment of a LDH sample with a Al/(Mg + Al) ratio equal to 0.33. Five main peaks, corresponding to the release of adsorbed molecules, were observed in the range of 298–1273 K. The first peak appeared below 473 K, and did not affect the integrity of the double layered structure. However the material was partially dehydrated and a significant rearrangement of the octahedral brucite-type layer occurred. The Al^{3+} cations migrated out of the layer to tetrahedral sites in the interlayer producing a decrease in the

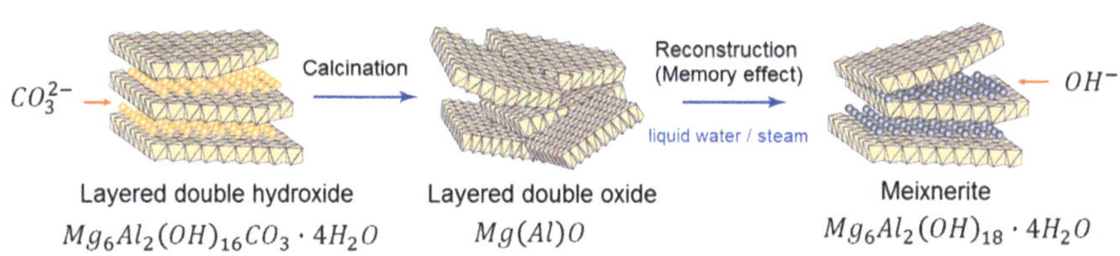

Fig. 2.4 Activation of LDH by thermal treatment followed by reconstruction (adapted from Tichit and Coq [71])

d spacings which progressed with temperature. The second peak was identified between 473 and 623 K. In this range of temperature the dehydroxylation of OH bonds with Al^{3+} occurred and the layered structures collapsed. The third peak became visible at ~623 K and corresponded to the maximum CO_2 evolution from the solid. At this temperature the material was completely dehydrated and partially dehydroxylated giving place to a high surface material (~200 $m^2\ g^{-1}$). The authors confirmed by IR analysis that at 773 K, the carbonates that were still present on the solid represented 20 % of the total original value. The remaining carbonates evolve in two overlapping peaks centred at 873 and 1173 K. The release of these carbonates coincides with the migration of Al^{3+} to MgO, which leads to the formation of spinel structures ($MgAl_2O_4$). This fact suggests that the carbonates are coordinated with Mg^{2+} cations and this coordination is affected by the Al^{3+} ions migration.

2.4.2 Factors Impacting on the Basicity of Layered Double Hydroxides

As mentioned before, layered double hydroxides have an ordered structure in which layers formed by divalent and trivalent cations are separated by anions and water molecules. These materials behave as solid bases and, depending on their composition and on the pretreatments that they are subjected to, a wide range of basic densities and strengths can be obtained. As-synthesised and reconstructed LDHs usually present very weak Brønsted basicity which is mainly dictated by the nature of the intercalated anion ($CO_3{}^{2-}$, NO^{3-}, Cl^-, $SO_4{}^{2-}$) and by the amount of remaining water. After calcination, the compensating ions and water are released and the resulting highly defective layered double oxides (LDOs) exhibit much higher (Lewis type) basicity than the fresh LDHs. In consequence, the basicity of the LDHs increases with the calcination temperature up to 923 K, when phase segregation and spinel ($MgAl_2O_4$) formation occur [63].

There are many factors that influence the basicity of layered double oxides. Although some weak basicity from remaining hydroxyls may exist in the calcined materials, the nature and the relative amount of the divalent and trivalent cations are the main factors that govern the basicity of the mixed oxides. In general, both metal ions must have radii close to 0.65 Å to form stable structures but small variations in the size produce significant differences in the strength of the basic sites [63]. Low charge-to-radius ratios of the cations result in higher negative charge densities located on the linking oxygen atoms rendering stronger bases. In the particular case of Al^{3+} and Mg^{2+}, aluminium oxide is an amphoteric material with higher heats of adsorption for NH_3 (~130 kJ mol^{-1}) than for CO_2 (~70 kJ mol^{-1}) while magnesium oxide chemisorbs CO_2 exclusively with an initial heat of adsorption of 120 kJ mol^{-1} [77]. Compared to the basicity of MgO, LDOs exhibit lower basic density but stronger sites which are attributed to the introduction of aluminium in the structure. Although the nature of this synergy is still a matter of discussion, it seems that the addition of aluminium increases the number of surface defects and tends to decrease the size of the LDO crystallites [71].

Carbon dioxide has enough acidity to chemisorb even on the weakest basic sites of the layered double oxides. Temperature-programmed desorption monitored by IR spectroscopy [78, 79] conducted over LDOs calcined at 773 K shows that CO_2 coordinates weakly with surface hydroxyl groups to form surface bicarbonate anions. Adsorption sites of intermediate strength correspond to bidentate carbonates coordinated to acid-base pairs such as Mg^{2+}–O^{2-} or Al^{3+}–O^{2-}. The strongest basic sites are monodentate carbonates that appear over low coordination (isolated) oxygen anions associated with defective surfaces [63, 80]. The adsorbed CO_2 species on LDOs are depicted in Fig. 2.5. The promotion of layered double oxides with alkali metals is a common strategy to tune their basicity for gas-phase adsorption applications. The presence of monovalent alkali ions such as sodium and potassium increases considerably the number of basic sites in the resulting LDO as well as the relative proportion of monodentate to bidentate carbonates [63].

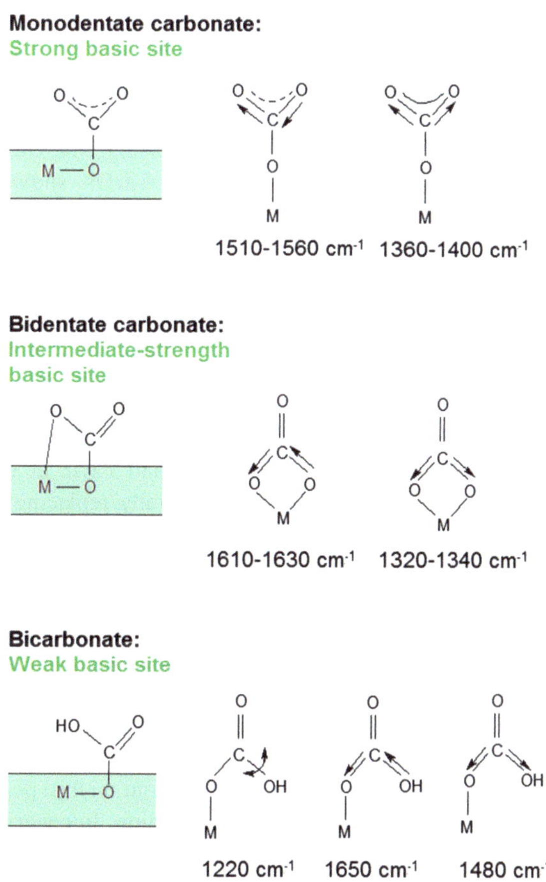

Fig. 2.5 Adsorbed CO_2 species on LDH and their corresponding IR signals (adapted from Di Cosimo et al. [79])

2.4.3 Strategies to Improve the Adsorption Performance of Layered Double Hydroxides

An adequate CO_2 adsorption performance requires high equilibrium capacities, fast kinetics and good regenerability. While layered double hydroxide derivatives (i.e. layered double oxides) exhibit outstanding kinetics compared to other materials that adsorb CO_2 in the same temperature range, their capacity and multicycle stability need to be further improved before technologies based on LDOs are industrially implemented. Different strategies have been explored to overcome these limitations including the substitution of framework metals or the promotion with alkali metals to increase their capacity, and the addition of water in the feed to enhance their capacity and regenerability. Additionally, it has been reported that dispersion on different supports increases the amount of CO_2 that can be adsorbed per mass of LDO. In the present section, the most common methods to enhance the adsorption performance of hydrotalcite derivatives are reviewed.

2.4.3.1 Substitution of Framework Metals

Numerous studies have attempted to increase the adsorption capacity of Mg–Al–CO_3 LDOs by substituting the structural ions. Tsuji et al. [81] synthesised a range of LDOs with different divalent cations (e.g. Mg, Cu, Zn, Ni and Co) and found that the Mg–Al–CO_3 and Co–Al–CO_3 LDOs showed the highest CO_2 uptake ($\sim$0.4 mol CO_2 kg^{-1} at 298 K and $p_{CO_2} = 66\,\text{mbar}$). Rodrigues et al. [75] investigated the effect of the type of compensating anion and concluded that lower adsorption capacities were obtained with OH^- compared to $CO_3{}^{2-}$ (0.1 and 0.2 mol CO_2 kg^{-1} respectively at 573 K and $p_{CO_2} = 50\,\text{mbar}$). Hutson et al. [18] assessed the CO_2 adsorption properties of Mg/Al and Ca/Al LDOs with different anions including $CO_3{}^{2-}$, $Fe(CN_6)^{4-}$, Cl^- and $ClO_4{}^-$. Among all the samples studied, a synthetic LDO with the formula of the natural mineral hydrotalcite exhibited the fastest kinetics and the highest adsorption capacity (0.6 mol CO_2 kg^{-1} at 573 K and $p_{CO_2} = 1000\,\text{mbar}$). White et al. [82] and Yavuz et al. [83] improved the adsorption capacity of LDOs by partially or totally replacing Al with Ga, La or Ce. More recently, Wang et al. [84] investigated the effect of different trivalent cations (Al, Fe, Ga and Mn) on the CO_2 adsorption performance of LDOs containing Mg^{2+} and $CO_3{}^{2-}$. The authors demonstrated that the type of trivalent cation used determines the structure evolution of LDH derivatives during thermal treatment and consequently influences their CO_2 capacity.

In addition to framework metal substitution, extensive research has been focused on increasing the CO_2 capacity of Mg–Al LDOs by varying the Mg^{2+} to Al^{3+} ratio [62, 75, 85, 86]. However, there has been little agreement regarding the optimum ratio which seems to be highly dependent on the specific conditions used during the thermal treatment and the CO_2 adsorption. In general, Mg/Al ratios in the range of 1 to 3 have been shown to be suitable for CO_2 adsorption at high temperature.

2.4.3.2 Alkali Promotion of Layered Double Hydroxides

The promotion of layered double hydroxides by alkaline dopants is one of the most practical ways to enhance their CO_2 adsorption capacity. As mentioned earlier, the presence of alkali ions (e.g. sodium, potassium and caesium) increases significantly the number and strength of basic sites in the LDHs. Alkali metals are often added to the LDHs by conventional impregnation methods but can also be incorporated by leaving residues from the coprecipitation agents (e.g. NaOH, KOH, Na_2CO_3 and K_2CO_3) [61, 87, 88].

Most of the research concerning alkali based LDHs has focused on investigating the adsorption properties of potassium promoted hydrotalcites and their application to sorption- assisted reactions such as the SEWGS and the SESMR [23, 54, 55, 85, 89–92]. Reijers et al. [85] assessed the CO_2 adsorption performance of a series of unpromoted and potassium promoted Mg–Al LDHs to be used in SESMR. The authors found that the CO_2 uptake at 673 and 773 K was significantly higher for the K-LDH samples but was practically insensitive to the potassium carbonate loading in the adsorbent, which was varied between 11 and 44 wt%. Rodrigues et al. [90] reported that the adsorption capacity of LDHs was considerably enhanced by promoting them with potassium and caesium. Since the surface area of the alkali loaded adsorbents was found to be much lower than that of the alkali free LDHs, they concluded that the increase in the CO_2 uptake was not directly related to the porosity of the sample but to the presence of additional interactions with or aided by the alkali cations. Walspurger et al. [91] investigated the type of active species responsible for the CO_2 adsorption on potassium promoted LDHs under conditions relevant to the SEWGS. They demonstrated that potassium ions strongly interact with the aluminium oxide centres in the LDH and this generates basic sites which reversibly adsorb CO_2 at 673 K.

2.4.3.3 CO_2 Adsorption in the Presence of Water

It has been widely reported that the presence of water has a positive influence on the CO_2 adsorption performance of activated layered double hydroxides. Ding et al. [45] studied the effect of steam on the adsorption and desorption steps of a pressure swing process using a potassium promoted LDO. They found that the adsorption capacities of the K-LDO at 673 and 753 K were $\sim$10 % higher under wet conditions compared to those obtained under dry conditions. Additionally, the multi-cycle stability of the adsorbent was significantly better in the presence of water. Remarkably, the partial pressure of water was found to have little effect on the adsorption capacity, which suggested that even low concentrations of steam are able to further activate the adsorption sites possibly by maintaining the hydroxyl concentration of the surface and/or preventing site poisoning through carbonate or coke deposition. In a subsequent work, Ding et al. [93] demonstrated that K-LDOs exhibit good adsorption performance even in the presence of considerably high concentrations of water (30 % v/v of water).

In a more recent study, Costa et al. [94] corroborated the findings published by Ding et al. showing that the CO_2 capacity at 473 K and the stability after TSA cycling of unpromoted LDOs were considerably enhanced by adding steam to the gas feed. The authors suggested that in the presence of carbon dioxide and water MgO and Al_2O_3 form bicarbonate species, $Mg(HCO_3)_2$ and $Al(HCO_3)_3$ which are responsible for the increased CO_2 uptake observed. Rodrigues et al. [75] reported that the presence of molecular water in the structure of the LDOs also favours their adsorption capacity. Similarly to Costa et al., they attributed the observed enhancement to the formation of bicarbonate species.

Regarding other competing species, it has been observed that K-promoted LDOs are able to coadsorb H_2S without affecting their CO_2 capacity significantly. It was proposed that the adsorption of sulphur occurs on two different types of sites: in some of them H_2S and CO_2 compete for adsorption although the affinity is higher for CO_2, while others adsorb H_2S exclusively, possibly as metal sulphides. Therefore, LDOs can be used in processes that involve the treatment of sour streams [95].

2.4.3.4 Supported Layered Double Hydroxides

Recently, it has been reported that the intrinsic capacity of activated LDHs (i.e. the amount of CO_2 adsorbed per mass of LDO) increases when they are dispersed over materials with high surface area. Although to date there are few works addressing this effect, it is expected that more active LDH derivatives will be obtained following this novel approach. The beneficial role of the support on the capacity of LDOs may stem from the stabilisation of smaller and more defective crystals, and from an increased accessibility of CO_2 to the adsorption sites. As mentioned above, the basicity of LDOs is a size-dependent characteristic since strong sites are associated with low coordinated O^{2-} anions which are present at the corners and the edges of the crystallites. Consequently, it is important to identify supports that are able to stabilise LDHs and that prevent sintering during multicycle operation.

Othman et al. [96] reported that zeolites coated with LDHs (Mg:Al ratio of 3) exhibit almost double the CO_2 adsorption capacity of uncoated LDOs in a temperature range from 303 to 573 K [62]. However, since the zeolites used are CO_2 adsorbents themselves, especially at low temperature, it is difficult to assess the influence of the support on the intrinsic capacity of the LDH derivatives. Aschenbrenner et al. [97] combined LDHs with bohemite alumina consisting of 50:50 LDH:alumina by weight and observed an enhancement of the thermal stability and adequate capacities (between 2 and 2.5 mol CO_2 kg^{-1} at 293 K), which were found to be independent of the divalent to trivalent cation ratio. The most notable improvement in the intrinsic capacity of LDOs (per mass of LDO) was reported by Meis and co-workers [80]. These authors found that the intrinsic capacity of unsupported LDOs was increased by an order of magnitude (1.3–2.5 mol CO_2 kg^{-1} at 523 K) when supported on a large fraction of carbon nanofibers (CNF) matrix (82–97 wt% CNF) and in the presence of water. They also showed that the addition of potassium to the CNF-supported materials results in even higher CO_2 capacities [88].

Although the use of supports appears to be effective for the improvement of the intrinsic sorption performance of LDOs, there are some drawbacks associated with this strategy. The main disadvantage is that so far, the increase in the intrinsic capacity of the layered oxides has been observed only when the crystallites are very well dispersed. This requires very high loadings of support which decreases the CO_2 capacity per total volume of adsorbent (LDO + support) and results in larger sorption units with higher overall costs. Consequently, the identification of supports that maximise the stability and the adsorption capacity of LDOs at relatively low loadings is crucial for industrial applications.

2.5 Carbon Nanostructures as Supports

The high surface area and porosity of some forms of carbon have awakened interest on their use as solid adsorbents or as supports in heterogeneous catalysis. Graphitic carbon exhibits a remarkable mechanical resistance and high stability in acidic and basic media. Additionally, the dispersion of the active phase is relatively easy compared to other supports. Until now there are few industrial processes that employ carbon as a support and only activated carbon or carbon blacks have been used so far [98]. However, nanostructured forms of carbon offer the possibility to improve diffusion due to larger intraparticle voids and therefore to fully exploit the high surface area of these materials [99].

Graphitic crystalline structures are obtained by combining carbon atoms having sp^2 hybrid orbitals. This configuration results in three σ-bonds localized along the plane connecting the carbon atoms which are arranged in a honeycomb 2D lattice known as *graphene*. The remaining $2p_z$ free electrons are weakly held to the nuclei and are distributed normal to the plane connecting the carbon atoms. The σ-covalent bonds are responsible for the great strength and mechanical properties of graphene while the delocalized $2p_z$ electrons are responsible for its electronic properties [100]. Two-dimensional graphene is considered to be the building block of several carbon allotropes. For instance, when many layers of graphene (typically more than three) are stacked one on top of the other, the three dimensional structure of *graphite* is obtained. Alternatively, graphene sheets can wrap forming hollow cylindrical structures known as *single-walled carbon nanotubes* (*SWCNTs*) with a diameter that ranges from about 0.5–5 nm and a specific surface area of 1315 $m^2 g^{-1}$, Fig. 2.6. When many graphene layers (from two to tens) arrange in concentric cylinders, *multi-walled carbon nanotubes (MWCNTs)* are obtained. MWCNTs have outer diameters in the range of 1.4–100 nm and surface areas between 700 and 800 $m^2 g^{-1}$ [101].

Graphene and carbon nanotubes are usually synthesised by chemical vapour deposition using Ni, Fe or Cu as catalysts. The nucleation and growth usually occurs by exposure of the transition metals to CO or to a hydrocarbon gas at

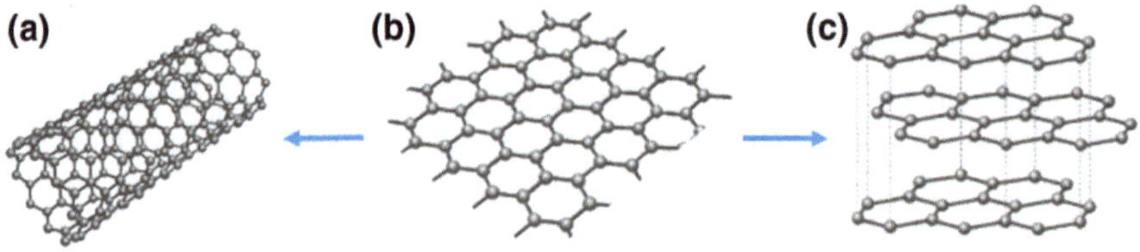

Fig. 2.6 Illustration of single-walled nanotube (**a**), graphene (**b**) and graphite (**c**)

temperatures around 1000 K and low pressures. Alternatively, graphene layers can be produced by oxidation of graphite followed by chemical and thermal exfoliation to produce *graphene oxide* sheets [102]. In graphene oxide, part of the carbon atoms change hybridization to sp^3 to combine with hydroxyls and peroxides in the basal planes, and with carbonyl and carboxylic groups that are located at the edges [103]. An important property of graphene oxide is that it can be easily dispersed in water and other organic solvents, which makes it an attractive support for various materials such as polymers and ceramics. The as-synthesised graphene, graphene oxide and CNTs are often contaminated with residues of the metal precursors and with undesirable by-products such as amorphous carbon. To remove these impurities, the materials are often treated with oxidants such as HNO_3, H_2SO_4 and $KMnO_4$. This purification treatment introduces acidic functionalities (i.e. –COOH, –C = O, and –OH) which help to solubilise the carbon structures in aqueous media. Additionally, these attached oxidic groups play an important role in the synthesis of composite supported materials (adsorbents and catalysts), providing strong fiber-matrix bonding and thus improving their mechanical properties [101].

2.6 Perspectives on Hydrogen Production

Hydrogen has a wide range of applications in the chemical and petrochemical industries, being used to produce principally ammonia, refined petroleum products and a large variety of chemicals. In addition, it is used in the metallurgic, electronic, and pharmaceutical sectors. Excluding the aeronautic industry, hydrogen has rarely been used as a fuel [104]. However, recently it has been considered as a promising source of clean energy and therefore much research is focused on developing a hydrogen sustainable economy. Unlike fossil fuels, hydrogen burns cleanly, without emitting any environmental pollutants such as carbon dioxide.

Hydrogen is not a primary energy source, but rather an energy carrier. It must be first manufactured before it can be used as a fuel. Industrially, hydrogen is mainly produced from fossil fuels natural gas being the major contributor. Methane is converted to hydrogen by steam reforming (SR) or by partial oxidation with oxygen (PO), or by both in sequence, i.e. autothermal reforming (AR). Although partial oxidation and autothermal reforming are more efficient than simple steam reforming, the latter is preferred since it does not involve the presence of oxygen which gives rise to additional production costs [105].

The steam reforming process consists in the reaction of sulphur-free natural gas with steam over a nickel-based catalyst at high temperatures and pressures (1073–1173 K and 5–35 bar, respectively) [106]. The water gas shift reaction (WGS) also takes place to some extent and therefore the product stream contains important amounts of water gas (H_2, CO, CO_2, H_2O) and unreacted CH_4. Since these reactions are equilibrium limited, additional water gas shift reactors are placed after the reformer to increase the overall hydrogen yield. The first WGS reactor operates at 573–773 K whereas the second operates near 473 K. Typically, a chromium-iron oxide catalyst is used in the high temperature reactor and a copper-zinc catalyst in the low temperature reactor. Finally, different purification processes (e.g. pressure swing adsorption, amine scrubbing, methanation and preferential oxidation) are used to eliminate carbon dioxide, unreacted carbon monoxide and other impurities, Fig. 2.7. The selection of these purification processes depends on the final purity required [107].

Besides natural gas, hydrogen can be produced from coal by the so-called integrated gasification combined cycle (IGCC). This technology is very energy efficient allowing the co-generation of hydrogen and electricity. In this process, coal is gasified by partial oxidation with oxygen and steam at high temperature and pressure. The resulting synthesis gas (mainly CO and H_2, but also CO_2) is further reacted with steam to enhance the hydrogen yield by the WGS. Next, the gas is cleaned to recover hydrogen, Fig. 2.7. Regrettably, due to the low hydrogen to carbon ratio of coal, the amount of CO_2 emitted per unit of H_2 (or energy) is large compared to combined cycles based on methane or oil, and this has limited the implementation of coal processes. A current challenge is then to couple IGCC with systematic carbon dioxide capture and sequestration processes [104].

An innovative energy efficient technology to increase the hydrogen production in chemical industry has been recently patented by Haldor Topsøe [108]. This process, called methanol-to-shift, is carried out in a water gas shift reactor feeding

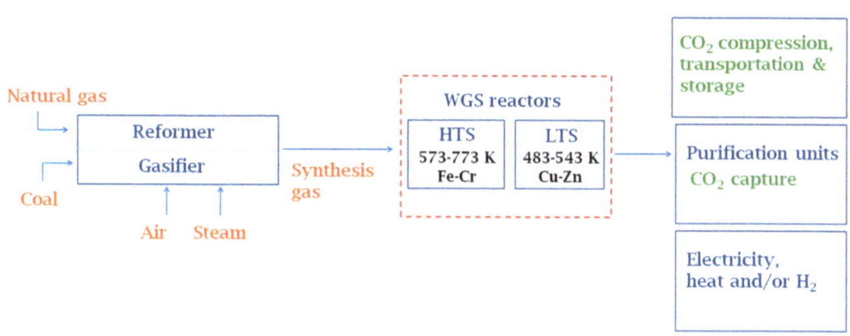

Fig. 2.7 Typical stages of the process of hydrogen production from natural gas (reforming) or coal (gasification)

simultaneously synthesis gas (mainly CO and H_2), water and methanol. A catalyst containing copper, zinc and aluminium enables hydrogen production from both CO shift conversion and methanol decomposition at operating conditions near 553 K and 20 bar. Further understanding of methanol-to-shift is needed since until now few studies have been published in the open literature [108, 109].

Small-scale hydrogen production can be performed via steam reforming, partial oxidation and autothermal reforming of hydrocarbons such as methane, methanol and ethanol. Although these technologies are more expensive and less efficient compared to large industrial processes, they avoid the cumbersome and dangerous transportation of hydrogen from distant centralized production centres. In addition, these processes can be coupled with small purification units to provide a hydrogen rich feed gas for fuel cells applications.

In recent decades new environmentally friendly processes, based on renewable sources rather than on fossil fuels, have been in continuous development to cover the increasing global demand of hydrogen. Important examples of these future technologies are the water electrolysis by alternative energies, photolysis (photo-electrochemical and photobiological) and biomass conversion. Regarding this last technology, biomass can be transformed into hydrogen by gasification or pyrolysis. In addition it can be used for the production of liquid fuels such as ethanol, methanol and glycerol (a major product of biodiesel), which subsequently can be transformed to hydrogen by steam reforming reactions [104, 105].

2.6.1 Sorption-Enhanced Hydrogen Production

Sorption-enhancement refers to intensified processes in which catalytic reactions are coupled with in situ separation by adsorption. These configurations may significantly improve the reactant conversion or/and the product selectivity. According to the Le Chatelier's principle, the removal of one of the products can shift the reaction equilibrium towards more favourable conditions than those achieved under conventional reaction operation. Adsorption-reaction systems enable lower operating temperatures reducing the energy demand of the process and preventing catalyst fouling (e.g. by carbon deposition). In addition, capital cost can be reduced due to process simplifications such as the use of less complicated purification units and the elimination of additional reactors used to enhance the conversion of an equilibrium limited reactor [107, 110].

In sorption-enhanced processes, the catalyst is usually physically admixed with the adsorbent and placed into the adsorptive reactor. Hence both solids and their relative quantities need to be carefully selected since adsorption and reaction kinetics must match under the same operating conditions. After the adsorbent is saturated it needs to be regenerated. If the regeneration conditions do not deactivate the catalyst, the adsorbent recovery can be carried out in two fixed-bed reactors, one operating in the adsorption-reaction mode and the other in the regeneration mode.

Otherwise, more complex engineering is required involving transported or fluidised systems [107].

Sorption-enhanced hydrogen production has been suggested to be feasible for methane steam reforming [53, 54, 107, 111–113], water gas shift reaction [55, 56, 95, 114–116], steam gasification of biomass [117, 118], and more recently for glycerol steam reforming [119, 120] and ethanol steam reforming [121–123]. In general the production of H_2 involves two primary reactions: the endothermic steam reforming and the moderately exothermic water gas shift (see Fig. 2.7). Since both reactions are thermodynamically limited, the in situ removal of carbon dioxide by a selective solid adsorbent displaces the reaction equilibrium and therefore almost complete conversion can be achieved. Although the main target of sorption-enhancement in the production of H_2 is to increase the amount of H_2 produced, the fact that CO_2 is removed simultaneously make the assisted process attractive as a pre-combustion CO_2 capture technology.

In the case of the sorption-assisted steam reforming process, more favourable thermodynamics allows for lower operating temperatures, which decrease the energy consumption and mediate the catalyst deactivation by coke formation. In addition, the water gas shift stage that usually follows the reformer may not be required, making the final purification of hydrogen simpler. Regrettably, a closer

Table 2.3 Sorption-enhanced processes for H_2 production

SE–H_2 process	Main reaction	Operating conditions (K)[a]	CO_2 adsorbents	Catalysts
Methane steam reforming	$CH_{4(g)} + H_2O_{(g)} \leftrightarrow 3H_{2(g)} + CO_{(g)}$	673–873	CaO [50, 107, 112] LDHs [53, 54, 85, 107, 111] Li_2ZrO_3 [107, 113] Na_2ZrO_3 [107, 113]	Ni based
Water gas shift reaction	$CO_{(g)} + H_2O_{(g)} \leftrightarrow H_{2(g)} + CO_{2(g)}$	573–873	CaO [115, 116] LDHs [55, 56, 95, 114]	Fe–Cr Pt-ZDC[b]
Steam gasification of biomass	$C_nH_mO_{p(g)} + (2n\text{-}p)H_2O_{(g)} \leftrightarrow (m/2 + 2n\text{-}p)H_{2(g)} + nCO_{2(g)}$	673–973	CaO [117, 118]	Ni based
Glycerol steam reforming	$C_3H_8O_{3(g)} + 3H_2O_{(g)} \leftrightarrow 7H_{2(g)} + 3CO_{2(g)}$	673–973	CaO [119, 120]	Ni based
Ethanol steam reforming	$C_2H_6O_{(g)} + 3H_2O_{(g)} \leftrightarrow 6H_{2(g)} + 2CO_{2(g)}$	773–973	CaO [121–123]	Ni based

[a]A wide range of pressures has been recommended, in general: 1–50 bar
[b]*ZDC* zirconia-doped ceria

analysis of the sorption-enhanced steam methane reforming (SESMR) reveals that in order to obtain adequate CH_4 conversions, extremely low CO_2 concentrations in the gas phase are required which are difficult to achieve in practice [124]. For instance, at 673 K and 1 bar, only 270 ppm of CO_2 are tolerated if conversions above 90 % are targeted, but the restrictions are even more severe if the process operates at higher pressures (10 ppb at 17 bar). On the other hand, the CO_2 adsorption requirements for the water gas shift are not so demanding and therefore the enhanced process is more feasible. Instead of having two shift reactors in series operating at high and low temperatures, sorption-enhanced water gas shift (SEWGS) requires only one reactor at intermediate temperature, where fast kinetics assisted by adsorption potentially results in smaller units.

Different chemisorbents such as calcium oxide, hydrotalcites and to a lesser extent lithium zirconates have been suggested as promising candidates for sorption-enhanced hydrogen production. Table 2.3 shows a summary of the mentioned hydrogen sorption-enhanced processes and the corresponding recommended adsorbents.

2.7 Concluding Remarks

The growing demand for ammonia derivatives and the need to process heavier petroleum feedstocks have generated great interest in the development of new processes for hydrogen production aiming to be more economic and environmentally friendly. In this regard, sorption-enhancement is a promising technology to improve the efficiency of processes in which H_2 is produced from conventional sources such as natural gas and coal, or from alternative feedstocks such as glycerol, ethanol and methanol. The main catalytic reactions involved (i.e. steam reforming and WGS) are thermodynamically limited and therefore the in situ removal of CO_2 by a selective solid adsorbent lifts the equilibrium limitations and maximises the hydrogen productivity. Sorption-assistance can be applied either in the reforming/gasification unit or in the water gas shift stage, although thermodynamic studies indicate the convenience of enhancing the latter.

Sorption-enhancement is also regarded as an attractive pre-combustion CO_2 capture technology. Therefore, extensive research has focused on finding adequate CO_2 solid adsorbents that operate under the relevant operating condition of these applications. Layered double hydroxides, also known as hydrotalcites, are one of the most promising groups of adsorbents for the enhancement of the water gas shift reaction, showing high selectivity towards CO_2 at temperatures between 473 and 723 K. In addition, they are sulphur tolerant and their activity is favoured by the presence of water. Compared to other solid adsorbents, LDHs show faster kinetics and the energy needed in their regeneration is considerably lower. Despite all these positive characteristics, LDHs show relatively low adsorption capacities and their multicycle stability needs to be improved before they are widely used in sorption-enhanced H_2 production and in other CCS processes.

Among the different efforts to enhance the performance of LDHs, emphasis has been given to the improvement of the adsorption capacity (e.g. framework substitution and alkali promotion) and only few studies have tackle the development of more stable adsorbents. A very novel strategy to increase the intrinsic capacity of hydrotalcites is the use of high surface area materials as support. It is expected that the resulting hybrids will exhibit better stability, although this effect has been seldom explored. This thesis focuses on the development of LDHs supported on carbon structures and on the assessment of their performance as CO_2 adsorbents. The aim of the present study is to obtain stable materials with an adequate capacity per mass of total adsorbent. Pure and alkali promoted adsorbents were tested. For some relevant samples, the adsorption thermodynamics and kinetics were studied to assess their suitability for sorption-enhanced hydrogen production processes.

References

1. Yang, R. T. (1997). *Gas separation by adsorption processes* (Vol. 1). London, UK: Imperial College Press.
2. Ponec, V., Knor, Z., & Cerny, S. (1974). *Adsorption on solids*. Prague, Czech Republic: London Butterworths.
3. Ruthven, D. M. (1984). *Principles of adsorption and adsorption processes*. NY, USA: Wiley.
4. Suzuki, M. (1990). *Adsorption engineering*. Tokyo, Japan: Kodansha.
5. Ruthven, D. M., Farooq, S., & Knaebel, K. S. (1994). *Pressure swing adsorption*. NY, USA: VCH Publishers.
6. Do, D. D. (1998). *Adsorption analysis: Equilibria and Kinetics*. London, UK: Imperial College Press.
7. Rouquerol, F., Rouquerol, J., & Sing, K. (1999). *Adsorption by powders & porous solids*. London, UK: Academic Press.
8. Keller, J. U., & Staudt, R. (2005). *Gas adsorption equilibria: Experimental methods and adsorption isotherms*. Boston, USA: Springer.
9. Smith, J. M., Van Ness, H. C., & Abbott, M. M. (2001). *Introduction to chemical engineering thermodynamics*. London, UK: McGraw-Hill.
10. Sircar, S., Mohr, R., Ristic, C., & Rao, M. B. (1999). Isosteric heat of adsorption: Theory and experiment. *The Journal of Physical Chemistry B, 103*(31), 6539–6546.
11. Attard, G., & Barnes, C. (2009). *Surfaces* (Vol. 59). NY, USA: Oxford University Press.
12. Koumpouras, G. (2007). Mathematical modelling of a low temperature hydrogen production process with in situ CO_2 capture. Ph.D. thesis, Imperial College London, London, UK.
13. Vannice, M. A. (2005). *Kinetics of catalytic reactions*. PA, USA: Springer.
14. Huang, T.-C., & Cho, L.-T. (1989). Relationships between constants of the Freundlich equation and temperature for gaseous adsorption. *Chemical Engineering Communications, 75*(1), 181–194.
15. Halsey, G., & Taylor, H. S. (1947). The adsorption of hydrogen on tungsten powders. *The Journal of Chemical Physics, 15*(9), 624–630.
16. Belmabkhout, Y., Frère, M., & Weireld, G. D. (2004). High-pressure adsorption measurements. A comparative study of the volumetric and gravimetric methods. *Measurement Science & Technology, 15*(5), 848.
17. Capezzuoli, F. (2006). Carbon dioxide adsorption on porous materials and its catalytic hydrogenation in supercritical phase. Ph.D. thesis, Imperial College London, London, UK.

18. Hutson, N. D., & Attwood, B. C. (2008). High temperature adsorption of CO_2 on various hydrotalcite-like compounds. *Adsorption, 14*(6), 781–789.
19. Du, H., Ebner, A. D., & Ritter, J. A. (2010). Temperature dependence of the nonequilibrium kinetic model that describes the adsorption and desorption behavior of CO_2 in K-promoted HTlc. *Industrial and Engineering Chemistry Research, 49*(7), 3328–3336.
20. Choi, S., Drese, J. H., & Jones, C. W. (2009). Adsorbent materials for carbon dioxide capture from large anthropogenic point sources. *ChemSusChem, 2*(9), 796–854.
21. Wood, G. O. (2002). Quantification and application of skew of breakthrough curves for gases and vapors eluting from activated carbon beds. *Carbon, 40*(11), 1883–1890.
22. Soares, J., Casarin, G., José, H., Moreira, R. P. M., & Rodrigues, A. (2005). Experimental and theoretical analysis for the CO_2 adsorption on hydrotalcite. *Adsorption, 11*(1), 237–241.
23. Lee, K. B., Verdooren, A., Caram, H. S., & Sircar, S. (2007). Chemisorption of carbon dioxide on potassium-carbonate-promoted hydrotalcite. *Journal of Colloid and Interface Science, 308*(1), 30–39.
24. Kolade, M. A. (2008). Adsorptive reactor technology for VOC abatement. PhD thesis, Imperial College London, London, UK.
25. Halmann, M. M., & Steinberg, M. (1999). *Greenhouse gas carbon dioxide mitigation: Science and technology*. Boca Raton, FL: Lewis Publishers.
26. Yong, Z., Mata, V., & Rodrigues, A. R. E. (2002). Adsorption of carbon dioxide at high temperature—a review. *Separation and Purification Technology, 26*(2–3), 195–205.
27. Siriwardane, R. V., Shen, M.-S., Fisher, E. P., & Losch, J. (2005). Adsorption of CO_2 on zeolites at moderate temperatures. *Energy & Fuels, 19*(3), 1153–1159.
28. Xiao, P., Zhang, J., Webley, P., Li, G., Singh, R., & Todd, R. (2008). Capture of CO_2 from flue gas streams with zeolite 13X by vacuum-pressure swing adsorption. *Adsorption, 14*(4–5), 575–582.
29. Chen, C., Park, D.-W., & Ahn, W.-S. (2014). CO_2 capture using zeolite 13X prepared from bentonite. *Applied Surface Science, 292*, 63–67.
30. Guo, B., Chang, L., & Xie, K. (2006). Adsorption of carbon dioxide on activated carbon. *Journal of Natural Gas Chemistry, 15*(3), 223–229.
31. Drage, T. C., Blackman, J. M., Pevida, C., & Snape, C. E. (2009). Evaluation of activated carbon adsorbents for CO_2 capture in gasification. *Energy & Fuels, 23*(5), 2790–2796.
32. Yin, G., Liu, Z., Liu, Q., & Wu, W. (2013). The role of different properties of activated carbon in CO_2 adsorption. *Chemical Engineering Journal, 230*, 133–140.
33. Watabe, T., Nishizaka, Y., Kazama, S., & Yogo, K. (2013). Development of amine-modified solid sorbents for postcombustion CO_2 capture. *Energy Procedia, 37*, 199–204.
34. Dutcher, B., Fan, M., Cui, S., Shen, X.-D., Kong, Y., Russell, A. G., et al. (2013). Characterization and stability of a new, high-capacity amine-functionalized CO_2 sorbent. *International Journal of Greenhouse Gas Control, 18*, 51–56.
35. Le, Y., Guo, D., Cheng, B., & Yu, J. (2013). Amine-functionalized monodispersed porous silica microspheres with enhanced CO_2 adsorption performance and good cyclic stability. *Journal of Colloid and Interface Science, 408*, 173–180.
36. Yazaydın, A. Ö., Benin, A. I., Faheem, S. A., Jakubczak, P., Low, J. J., Willis, R. R., & Snurr, R. Q. (2009). Enhanced CO_2 adsorption in metal-organic frameworks via occupation of open-metal sites by coordinated water molecules. *Chemistry of Materials, 21*(8), 1425–1430.
37. Li, J.-R., Ma, Y., McCarthy, M. C., Sculley, J., Yu, J., Jeong, H.-K., et al. (2011). Carbon dioxide capture-related gas adsorption and separation in metal-organic frameworks. *Coordination Chemistry Reviews, 255*(15–16), 1791–1823.
38. Ye, S., Jiang, X., Ruan, L.-W., Liu, B., Wang, Y.-M., Zhu, J.-F., & Qiu, L.-G. (2013). Post-combustion CO_2 capture with the HKUST-1 and MIL-101(Cr) metal–organic frameworks: Adsorption, separation and regeneration investigations. *Microporous and Mesoporous Materials, 179*, 191–197.

39. Yong, Z., Mata, V., & Rodrigues, A. (2001). Adsorption of carbon dioxide on chemically modified high surface area carbon-based adsorbents at high temperature. *Adsorption, 7*(1), 41–50.
40. Xiao, P., Wilson, S., Xiao, G., Singh, R., & Webley, P. (2009). Novel adsorption processes for carbon dioxide capture within a IGCC process. *Energy Procedia, 1*(1), 631–638.
41. Lee, S. C., Chae, H. J., Lee, S. J., Choi, B. Y., Yi, C. K., Lee, J. B., et al. (2008). Development of regenerable MgO-based sorbent promoted with K_2CO_3 for CO_2 capture at low temperatures. *Environmental Science and Technology, 42*(8), 2736–2741.
42. Bhagiyalakshmi, M., Lee, J. Y., & Jang, H. T. (2010). Synthesis of mesoporous magnesium oxide: Its application to CO_2 chemisorption. *International Journal of Greenhouse Gas Control, 4*(1), 51–56.
43. Lee, K. B., Beaver, M. G., Caram, H. S., & Sircar, S. (2007). Chemisorption of carbon dioxide on sodium oxide promoted alumina. *AIChE Journal, 53*(11), 2824–2831.
44. Beaver, M., & Sircar, S. (2010). Adsorption technology for direct recovery of compressed, pure CO_2 from a flue gas without pre-compression or pre-drying. *Adsorption, 16*(3), 103–111.
45. Ding, Y., & Alpay, E. (2000). Equilibria and kinetics of CO_2 adsorption on hydrotalcite adsorbent. *Chemical Engineering Science, 55*(17), 3461–3474.
46. Yong, Z., & Rodrigues, A. R. E. (2002) Hydrotalcite-like compounds as adsorbents for carbon dioxide. *Energy Conversion and Management, 43*(14), 1865–1876.
47. Xiong, R., Ida, J., & Lin, Y. S. (2003). Kinetics of carbon dioxide sorption on potassium-doped lithium zirconate. *Chemical Engineering Science, 58*(19), 4377–4385.
48. Ochoa-Fernandez, E., Ronning, M., Yu, X., Grande, T., & Chen, D. (2007). Compositional effects of nanocrystalline lithium zirconate on its CO_2 capture properties. *Industrial and Engineering Chemistry Research, 47*(2), 434–442.
49. Iwan, A., Stephenson, H., Ketchie, W. C., & Lapkin, A. A. (2009). High temperature sequestration of CO_2 using lithium zirconates. *Chemical Engineering Journal, 146*(2), 249–258.
50. Solieman, A. A. A., Dijkstra, J. W., Haije, W. G., Cobden, P. D., & van den Brink, R. W. (2009). Calcium oxide for CO_2 capture: Operational window and efficiency penalty in sorption-enhanced steam methane reforming. *International Journal of Greenhouse Gas Control, 3*(4), 393–400.
51. Dou, B., Song, Y., Liu, Y., & Feng, C. (2010). High temperature CO_2 capture using calcium oxide sorbent in a fixed-bed reactor. *Journal of Hazardous Materials, 183*(1–3), 759–765.
52. Dean, C. C., Blamey, J., Florin, N. H., Al-Jeboori, M. J., & Fennell, P. S. (2011). The calcium looping cycle for CO_2 capture from power generation, cement manufacture and hydrogen production. *Chemical Engineering Research and Design, 89*(6), 836–855.
53. Hufton, J. R., Mayorga, S., & Sircar, S. (1999). Sorption-enhanced reaction process for hydrogen production. *AIChE Journal, 45*(2), 248–256.
54. Ding, Y., & Alpay, E. (2000). Adsorption-enhanced steam–methane reforming. *Chemical Engineering Science, 55*(18), 3929–3940.
55. van Selow, E. R., Cobden, P. D., Verbraeken, P. A., Hufton, J. R., & van den Brink, R. W. (2009). Carbon capture by sorption-enhanced water–gas shift reaction process using hydrotalcite-based material. *Industrial and Engineering Chemistry Research, 48*(9), 4184–4193.
56. Jang, H. M., Lee, K. B., Caram, H. S., & Sircar, S. (2012). High-purity hydrogen production through sorption enhanced water gas shift reaction using K_2CO_3-promoted hydrotalcite. *Chemical Engineering Science, 73*, 431–438.
57. Cavani, F., Trifirò, F., & Vaccari, A. (1991). Hydrotalcite-type anionic clays: Preparation, properties and applications. *Catalysis Today, 11*(2), 173–301.
58. Rousselot, I., Taviot-Guého, C., Leroux, F., Léone, P., Palvadeau, P., & Besse, J.-P. (2002). Insights on the structural chemistry of hydrocalumite and hydrotalcite-like materials: Investigation of the series $Ca_2M^{3+}(OH)_6Cl{\cdot}2H_2O$ (M^{3+}: Al^{3+}, Ga^{3+}, Fe^{3+}, and Sc^{3+}) by X-ray powder diffraction. *Journal of Solid State Chemistry, 167*(1), 137–144.

59. He, J., Wei, M., Li, B., Kang, Y., Evans, D. G., & Duan, X. (2006). *Layered double hydroxides* (Vol. 119). Heidelberg, Germany: Springer.
60. Othman, M. R., Helwani, Z., & Fernando, W. J. N. (2009). Synthetic hydrotalcites from different routes and their application as catalysts and gas adsorbents: A review. *Applied Organometallic Chemistry, 23*(9), 335–346.
61. León, M., Díaz, E., Bennici, S., Vega, A., Ordóñez, S., & Auroux, A. (2010). Adsorption of CO_2 on hydrotalcite-derived mixed oxides: Sorption mechanisms and consequences for adsorption irreversibility. *Industrial and Engineering Chemistry Research, 49*(8), 3663–3671.
62. Gao, Y., Zhang, Z., Wu, J., Yi, X., Zheng, A., Umar, A., et al. (2013). Comprehensive investigation of CO_2 adsorption on Mg-Al-CO_3 LDH-derived mixed metal oxides. *Journal of Materials Chemistry A, 1*(41), 12782–12790.
63. Debecker, D. P., Gaigneaux, E. M., & Busca, G. (2009). Exploring, tuning, and exploiting the basicity of hydrotalcites for applications in heterogeneous catalysis. *Chemistry—A European Journal, 15*(16), 3920–3935.
64. Ebner, A. D., Reynolds, S. P., & Ritter, J. A. (2006). Understanding the adsorption and desorption behavior of CO_2 on a K-promoted hydrotalcite-like compound (HTlc) through nonequilibrium dynamic isotherms. *Industrial and Engineering Chemistry Research, 45*(18), 6387–6392.
65. Ram Reddy, M. K., Xu, Z. P., Lu, G. Q., & Diniz da Costa, J. C. (2006). Layered double hydroxides for CO_2 capture: Structure evolution and regeneration. *Industrial & Engineering Chemistry Research, 45*(22), 7504–7509.
66. Wang, X. P., Yu, J. J., Cheng, J., Hao, Z. P., & Xu, Z. P. (2007). High-temperature adsorption of carbon dioxide on mixed oxides derived from hydrotalcite-like compounds. *Environmental Science and Technology, 42*(2), 614–618.
67. Hutson, N. D., Speakman, S. A., & Payzant, E. A. (2004). Structural effects on the high temperature adsorption of CO_2 on a synthetic hydrotalcite. *Chemistry of Materials, 16*(21), 4135–4143.
68. Turco, M., Bagnasco, G., Costantino, U., Marmottini, F., Montanari, T., Ramis, G., & Busca, G. (2004). Production of hydrogen from oxidative steam reforming of methanol: I. Preparation and characterization of Cu/ZnO/Al_2O_3 catalysts from a hydrotalcite-like LDH precursor. *Journal of Catalysis, 228*(1), 43–55.
69. Pérez-Ramírez, J., Abelló, S., & van der Pers, N. M. (2007). Memory effect of activated Mg–Al hydrotalcite: In Situ XRD studies during decomposition and gas-phase reconstruction. *Chemistry—A European Journal, 13*(3), 870–878.
70. Di Cosimo, J. I., Díez, V. K., & Apesteguía, C. R. (1996). Base catalysis for the synthesis of α, β-unsaturated ketones from the vapor-phase aldol condensation of acetone. *Applied Catalysis, A: General, 137*(1), 149–166.
71. Tichit, D., & Coq, B. (2003). Catalysis by hydrotalcites and related materials. *CATTECH, 7*(6), 206–217.
72. Climent, M. J., Corma, A., Iborra, S., Epping, K., & Velty, A. (2004). Increasing the basicity and catalytic activity of hydrotalcites by different synthesis procedures. *Journal of Catalysis, 225*(2), 316–326.
73. Abelló, S., Medina, F., Tichit, D., Pérez-Ramírez, J., Groen, J. C., Sueiras, J. E., et al. (2005). Aldol condensations over reconstructed Mg–Al hydrotalcites: Structure-activity relationships related to the rehydration method. *Chemistry—A European Journal, 11*(2), 728–739.
74. Hibino, T., Yamashita, Y., Kosuge, K., & Tsunashima, A. (1995). Decarbonation behavior of Mg-Al-CO 3 hydrotalcite-like compounds during heat treatment. *Clays and Clay Minerals, 43*(4), 427–432.
75. Yong, Z., Mata, V., & Rodrigues, A. E. (2000). Adsorption of carbon dioxide onto hydrotalcite-like compounds (HTlcs) at high temperatures. *Industrial and Engineering Chemistry Research, 40*(1), 204–209.

76. Yang, W., Kim, Y., Liu, P. K. T., Sahimi, M., & Tsotsis, T. T. (2002). A study by in situ techniques of the thermal evolution of the structure of a Mg–Al–CO_3 layered double hydroxide. *Chemical Engineering Science, 57*(15), 2945–2953.
77. Auroux, A., & Gervasini, A. (1990). Microcalorimetric study of the acidity and basicity of metal oxide surfaces. *The Journal of Physical Chemistry, 94*(16), 6371–6379.
78. Philipp, R., & Fujimoto, K. (1992). FTIR spectroscopic study of carbon dioxide adsorption/desorption on magnesia/calcium oxide catalysts. *The Journal of Physical Chemistry, 96*(22), 9035–9038.
79. Di Cosimo, J. I., Díez, V. K., Xu, M., Iglesia, E., & Apesteguía, C. R. (1998). Structure and surface and catalytic properties of Mg-Al basic oxides. *Journal of Catalysis, 178*(2), 499–510.
80. Meis, N. N. A. H., Bitter, J. H., & de Jong, K. P. (2009). Support and size effects of activated hydrotalcites for precombustion CO_2 capture. *Industrial and Engineering Chemistry Research, 49*(3), 1229–1235.
81. Tsuji, M., Mao, G., Yoshida, T., & Tamaura, Y. (1993). Hydrotalcites with an extended Al^{3+}-substitution: Synthesis, simultaneous TG-DTA-MS study, and their CO_2 adsorption behaviors. *Journal of Materials Research, 8*(05), 1137–1142.
82. White, M., Iretski, A., Weigel, S., Chiang, R., & Brozozowski, J. (2006). *Adsorbents, methods of preparation, and methods of use thereof.* Google Patents.
83. Yavuz, C. T., Shinall, B. D., Iretskii, A. V., White, M. G., Golden, T., Atilhan, M., et al. (2009). Markedly improved CO_2 capture efficiency and stability of gallium substituted hydrotalcites at elevated temperatures. *Chemistry of Materials, 21*(15), 3473–3475.
84. Wang, Q., Tay, H. H., Ng, D. J. W., Chen, L., Liu, Y., Chang, J., et al. (2010). The effect of trivalent cations on the performance of Mg–M–CO_3 layered double hydroxides for high-temperature CO_2 capture. *ChemSusChem, 3*(8), 965–973.
85. Reijers, H. T. J., Valster-Schiermeier, S. E. A., Cobden, P. D., & van den Brink, R. W. (2005). Hydrotalcite as CO_2 sorbent for sorption-enhanced steam reforming of methane. *Industrial and Engineering Chemistry Research, 45*(8), 2522–2530.
86. Yang, J.-I., & Kim, J.-N. (2006). Hydrotalcites for adsorption of CO_2 at high temperature. *Korean Journal of Chemical Engineering, 23*(1), 77–80.
87. Abelló, S., Medina, F., Tichit, D., Pérez-Ramírez, J., Rodríguez, X., Sueiras, J. E., et al. (2005). Study of alkaline-doping agents on the performance of reconstructed Mg–Al hydrotalcites in aldol condensations. *Applied Catalysis, A: General, 281*(1–2), 191–198.
88. Meis, N. N. A. H., Bitter, J. H., & de Jong, K. P. (2010). On the influence and role of alkali metals on supported and unsupported activated hydrotalcites for CO_2 sorption. *Industrial and Engineering Chemistry Research, 49*(17), 8086–8093.
89. Ebner, A. D., Reynolds, S. P., & Ritter, J. A. (2007). Nonequilibrium kinetic model that describes the reversible adsorption and desorption behavior of CO_2 in a K-promoted hydrotalcite-like compound. *Industrial and Engineering Chemistry Research, 46*(6), 1737–1744.
90. Oliveira, E. L. G., Grande, C. A., & Rodrigues, A. E. (2008). CO_2 sorption on hydrotalcite and alkali-modified (K and Cs) hydrotalcites at high temperatures. *Separation and Purification Technology, 62*(1), 137–147.
91. Walspurger, S., Boels, L., Cobden, P. D., Elzinga, G. D., Haije, W. G., & van den Brink, R. W. (2008). The crucial role of the K^+–aluminium oxide interaction in K^+-promoted alumina- and hydrotalcite-based materials for CO_2 sorption at high temperatures. *ChemSusChem, 1*(7), 643–650.
92. Wu, Y. J., Li, P., Yu, J. G., Cunha, A. F., & Rodrigues, A. E. (2013). K-Promoted hydrotalcites for CO_2 capture in sorption enhanced reactions. *Chemical Engineering and Technology, 36*(4), 567–574.
93. Ding, Y., & Alpay, E. (2001). High temperature recovery of CO_2 from flue gases using hydrotalcite adsorbent. *Process Safety and Environmental Protection, 79*(1), 45–51.

94. Ram Reddy, M. K., Xu, Z. P., & Diniz da Costa, J. C. (2008). Influence of water on high-temperature CO_2 capture using layered double hydroxide derivatives. *Industrial & Engineering Chemistry Research, 47*(8), 2630–2635.
95. van Dijk, H. A. J., Walspurger, S., Cobden, P. D., van den Brink, R. W., & de Vos, F. G. (2011). Testing of hydrotalcite-based sorbents for CO_2 and H_2S capture for use in sorption enhanced water gas shift. *International Journal of Greenhouse Gas Control, 5*(3), 505–511.
96. Othman, M. R., Rasid, N. M., & Fernando, W. J. N. (2006). Mg–Al hydrotalcite coating on zeolites for improved carbon dioxide adsorption. *Chemical Engineering Science, 61*(5), 1555–1560.
97. Aschenbrenner, O., McGuire, P., Alsamaq, S., Wang, J., Supasitmongkol, S., Al-Duri, B., et al. (2011). Adsorption of carbon dioxide on hydrotalcite-like compounds of different compositions. *Chemical Engineering Research and Design, 89*(9), 1711–1721.
98. Tessonnier, J. P. (2009). Carbon materials in heterogeneous catalysis. In Planck, F.-M. (Ed,), *Lecture series: Modern methods in heterogeneous catalysis.*
99. Chinthaginjala, J. K., Seshan, K., & Lefferts, L. (2007). Preparation and application of carbon-nanofiber based microstructured materials as catalyst supports. *Industrial and Engineering Chemistry Research, 46*(12), 3968–3978.
100. Wong, H. S. P., & Akinwande, D. (2011). *Carbon nanotube and graphene device physics.* NY, USA: Cambridge University Press.
101. Rakov, E. G. (2006). *Nanotubes and nanofibers.* FL, USA: Taylor and Francis Group.
102. Machado, B. F., & Serp, P. (2012). Graphene-based materials for catalysis. *Catalysis Science & Technology, 2*(1), 54–75.
103. Lerf, A., He, H., Forster, M., & Klinowski, J. (1998). Structure of graphite oxide revisited. *The Journal of Physical Chemistry B, 102*(23), 4477–4482.
104. Olah, G. A., Goeppert, A., & Prakash, G. K. S. (2006). *Beyond oil and gas: The methanol economy.* Weinheim, Germany: Wiley-VCH.
105. Baade, W. F., Parekh, U. N., & Raman, V. S. (2001). *Hydrogen* (Vol. 13, pp. 759–808). Kirk-Othmer Encyclopedia (online).
106. Rostrup-Nielsen, J. R. (2008). Steam reforming. In *Handbook of heterogeneous catalysis.* Weinheim: Wiley-VCH Verlag GmbH & Co. KGaA.
107. Harrison, D. P. (2008). Sorption-enhanced hydrogen production: A review. *Industrial and Engineering Chemistry Research, 47*(17), 6486–6501.
108. Nielsen, P. E. H., Hansen, J. B., & Schiødt, N. C. (2009). Process for the preparation of a hydrogen-rich stream. US 7,527,781 B2.
109. Choi, Y., & Stenger, H. (2002). Kinetics of methanol decomposition and water gas shift reaction on a commercial Cu-ZnO/Al_2O_3 catalyst. *Fuel Chemistry Division Preprints, 47*, 723–724.
110. Carvill, B. T., Hufton, J. R., Anand, M., & Sircar, S. (1996). Sorption-enhanced reaction process. *AIChE Journal, 42*(10), 2765–2772.
111. Reijers, H. T. J., Boon, J., Elzinga, G. D., Cobden, P. D., Haije, W. G., & Brink, R. W. V. D. (2009). Modeling study of the sorption-enhanced reaction process for CO_2 capture. II. Application to steam-methane reforming. *Industrial & Engineering Chemistry Research, 48*(15), 6975–6982.
112. Broda, M., Manovic, V., Imtiaz, Q., Kierzkowska, A. M., Anthony, E. J., & Müller, C. R. (2013). High-purity hydrogen via the sorption-enhanced steam methane reforming reaction over a synthetic CaO-based sorbent and a Ni catalyst. *Environmental Science and Technology, 47*(11), 6007–6014.
113. Ochoa-Fernandez, E., Haugen, G., Zhao, T., Ronning, M., Aartun, I., Borresen, B., et al. (2007). Process design simulation of H_2 production by sorption enhanced steam methane reforming: evaluation of potential CO_2 acceptors. *Green Chemistry, 9*(6), 654–662.
114. van Selow, E. R., Cobden, P. D., van den Brink, R. W., Hufton, J. R., & Wright, A. (2009). Performance of sorption-enhanced water-gas shift as a pre-combustion CO_2 capture technology. *Energy Procedia, 1*(1), 689–696.

115. Stevens, R. W, Jr., Shamsi, A., Carpenter, S., & Siriwardane, R. (2010). Sorption-enhanced water gas shift reaction by sodium-promoted calcium oxides. *Fuel, 89*(6), 1280–1286.
116. Liu, Y., Li, Z., Xu, L., & Cai, N. (2012). Effect of sorbent type on the sorption enhanced water gas shift process in a fluidized bed reactor. *Industrial and Engineering Chemistry Research, 51*(37), 11989–11997.
117. Florin, N. H., & Harris, A. T. (2008). Enhanced hydrogen production from biomass with in situ carbon dioxide capture using calcium oxide sorbents. *Chemical Engineering Science, 63*(2), 287–316.
118. Pimenidou, P., Rickett, G., Dupont, V., & Twigg, M. V. (2010). High purity H_2 by sorption-enhanced chemical looping reforming of waste cooking oil in a packed bed reactor. *Bioresource technology, 101*(23), 9279–9286.
119. Chen, H., Zhang, T., Dou, B., Dupont, V., Williams, P., Ghadiri, M., & Ding, Y. (2009). Thermodynamic analyses of adsorption-enhanced steam reforming of glycerol for hydrogen production. *International Journal of Hydrogen Energy, 34*(17), 7208–7222.
120. Dou, B., Rickett, G. L., Dupont, V., Williams, P. T., Chen, H., Ding, Y., & Ghadiri, M. (2010). Steam reforming of crude glycerol with in situ CO_2 sorption. *Bioresource technology, 101*(7), 2436–2442.
121. Comas, J., Laborde, M., & Amadeo, N. (2004). Thermodynamic analysis of hydrogen production from ethanol using CaO as a CO_2 sorbent. *Journal of Power Sources, 138*(1–2), 61–67.
122. He, L., Berntsen, H., & Chen, D. (2009). Approaching sustainable H_2 production: Sorption enhanced steam reforming of ethanol. *The Journal of Physical Chemistry A, 114*(11), 3834–3844.
123. Li, M. (2009). Thermodynamic analysis of adsorption enhanced reforming of ethanol. *International Journal of Hydrogen Energy, 34*(23), 9362–9372.
124. Cobden, P. D., van Beurden, P., Reijers, H. T. J., Elzinga, G. D., Kluiters, S. C. A., Dijkstra, J. W., et al. (2007). Sorption-enhanced hydrogen production for pre-combustion CO_2 capture: Thermodynamic analysis and experimental results. *International Journal of Greenhouse Gas Control, 1*(2), 170–179.

Chapter 3
Experimental Methods

3.1 Introduction

In this chapter, the standard experimental procedures and the instruments used to test and characterise a wide range of LDH-based adsorbents and water-gas shift catalysts are presented. For the sake of clarity, the description of commercial materials and the methodology followed to synthesise and activate adsorbents and catalysts are detailed in the corresponding chapters. In Sect. 3.2 the use of a thermogravimetric analyser and the design and operation of a lab scale adsorption column are explained. These instruments were used to measure the CO_2 capacities and to assess the multicycle stability of different adsorbents. Additionally, the output of breakthrough curve experiments in the packed column allowed the kinetics of adsorption to be obtained. Section 3.3 includes a description of the experimental setup used to conduct a preliminary screening of catalysts for water-gas shift and methanol decomposition. The methodology followed to analyse the reaction products is explained. Finally, the characterisation techniques and the equipment used to determine the physicochemical properties of the adsorbents and catalysts studied are presented in Sect. 3.4.

3.2 Experimental Methods to Measure Adsorption

As part of the research, it was necessary to select, design and build experimental systems to study the CO_2 adsorption performance of different adsorbents. Considering the availability of equipment and the fact that dynamic techniques allow equilibrium and kinetic information to be obtained, thermogravimetric [1–3] and breakthrough curve analyses [4–6] were adopted as main methodologies (see Chap. 2, Sect. 2.2.6) [7, 8]. Thermal gravimetric analysis (TGA) was extensively used to determine the dry adsorption capacity and multicycle stability of solids that

D. Iruretagoyena Ferrer, *Supported Layered Double Hydroxides as CO_2 Adsorbents for Sorption-enhanced H_2 Production*, Springer Theses,
DOI 10.1007/978-3-319-41276-4_3

were often available in small amounts. On the other hand, an adsorption column was commissioned to measure the performance, under dry and wet conditions, of relevant adsorbents that were available in relatively larger amounts.

3.2.1 *Thermogravimetric Analysis*

3.2.1.1 Experimental Setup and Standard Procedure

Two different thermogravimetric analysers (TA Instruments, TAQ500 and Perkin Elmer, TGA4000) with an accuracy of ≤±0.1 % or 10 μg (whichever is greater) were used to measure the CO_2 capacity, stability and isotherms of different adsorbents. Prior to the CO_2 adsorption measurements, the samples were activated either in situ or ex situ at 673 K for 4 h under flowing N_2. The materials that were calcined ex situ were further pretreated at 673 K for 1 h flowing N_2 in the TGA instrument to remove CO_2 and moisture captured from the atmosphere during their transportation and storage. The amount of adsorbent used in the CO_2-TGA experiments was relatively small varying from 5 to 25 mg and the samples were very fine particles (powder) in all cases. After the adsorbent was activated/pretreated, it was cooled down to the selected adsorption temperature (typically 573 K) and held for 5–10 min The feed was then switched to a CO_2/N_2 or CO_2/Ar premixed gas and held until the weight signal was nearly steady. The adsorption capacity (i.e. working capacity[1]) of the solid was determined from the change in mass during CO_2 flow. The influence of the change in viscosity and density between N_2 and the CO_2 containing mixtures was corrected by measuring the response to a non-adsorbing material (aluminium foil or silicon carbide). This blank, which was commonly very small, was subtracted from the sample response.

The regeneration and stability of the adsorbents was assessed by multicycle tests in which the adsorption step was typically carried out flowing a premixed 20 % CO_2/N_2 gas (BOC) for 60 min, whereas the regeneration was conducted at 673 K for 30 min under flowing nitrogen. In addition, adsorption isotherms at 573 and 623 K were obtained using CO_2 partial pressures of 20, 100, 150, 200, 500, 800 and 990 mbar (premixed gases, balance Ar or N_2). First-contact adsorption isotherms were determined from a series of experiments using different CO_2 compositions and activating a fresh sample before each measurement. Adsorption capacities measured during the first exposure to the adsorptive gas were used in the plots. Alternatively, isotherms were obtained using samples that were subjected to multicycle tests. Continuous adsorption/desorption cycles were carried out (using 20 % CO_2/N_2 or 15 % CO_2/N_2 mixtures) until a constant capacity was observed.

[1]The adsorption capacities reported are working capacities rather than equilibrium values since longer equilibration times were not practical for the studies conducted. However, it is expected that the capacities measured are close to the thermodynamic values (see Sect. 3.2.1.2).

Table 3.1 Effect of the gas space velocity on the CO_2 uptake of an activated LDH sample (Mg/Al 2:1) at different times on stream

Time (min)	Adsorption capacity (mol CO_2 kg^{-1})[a]				
	Space velocity[b] $1.7\ \frac{mL}{min\ mg}$	Space velocity[c] $2.4\ \frac{mL}{min\ mg}$	Space velocity[d] $5.1\ \frac{mL}{min\ mg}$	Space velocity[e] $12.2\ \frac{mL}{min\ mg}$	Space velocity[f] $1.6\ \frac{mL}{min\ mg}$
1	0.16	0.15	0.14	0.15	0.15
3	0.20	0.20	0.19	0.19	0.20
5	0.22	0.22	0.21	0.20	0.21
30	0.26	0.26	0.26	0.25	0.25
60	0.27	0.28	0.27	0.27	0.26
120	0.28	0.29	0.28	0.29	0.28
600	0.31	0.32	0.31	0.31	0.30

[a]The LDH sample was synthesised by co-precipitation and activated ex situ at 673 K for 4 h by flowing 100 mL min^{-1} N_2
[b]TAQ500; Flow STP 40 mL min^{-1}; 573 K; PCO_2 = 200 mbar
[c]TAQ500; Flow STP 60 mL min^{-1}; 573 K; PCO_2 = 200 mbar
[d]TAQ500; Flow STP 60 mL min^{-1}; 573 K; PCO_2 = 200 mbar
[e]TAQ500; Flow STP 60 mL min^{-1}; 573 K; PCO_2 = 200 mbar
[f]TGA 4000; Flow STP 20 mL min^{-1}; 573 K; PCO_2 = 200 mbar

Then the CO_2 content in the adsorptive gas was decreased to 2 % CO_2/Ar and successively increased for the following adsorption steps. It is worth mentioning that in all the CO_2-TGA experiments the selected flow rate (20–60 mL min^{-1}) was kept constant during the adsorption and desorption steps.

3.2.1.2 Mass Transfer Effects

The fact that in the thermogravimetric analyser the gas flows over and not through the solid samples may result in the formation of a thin stagnant film through which the gas molecules in the bulk have to diffuse to reach the surface of the adsorbent. In addition, the size and shape of the pores of the adsorbent may restrict the transport of molecules from the surface to the sites located in the interior of the particles.

To assess the influence of the external mass transfer limitations in the TGA measurements, the CO_2 adsorption capacity of a pure LDH (Mg/Al = 2) sample was determined for a fixed particle size and different gas space velocities,[2] Table 3.1. It can be observed that, in the range of masses and flows evaluated, the adsorption capacities obtained at different times on stream are the same within the uncertainty of the apparatus. Moreover, a comparison between the results obtained

[2]The space velocity is defined as volumetric flow rate of the CO_2 premixed gas divided by the mass of adsorbent.

Table 3.2 Effect of particle size on the CO_2 uptake of an activated LDH sample (Mg/Al 2:1) at different times on stream

Time (min)	Adsorption capacity (mol CO_2 kg^{-1})[a]	
	Fine particles (10 μm)[b]	250–500 μm[c]
1	0.14	0.10
3	0.19	0.15
5	0.21	0.18
30	0.26	0.24
60	0.27	0.26
120	0.28	0.27
600	0.31	0.29

[a]The LDH sample was synthesised by co-precipitation and was activated ex situ at 673 K for 4 h by flowing 100 mL min^{-1} N_2
[b]TAQ500; SV 5.1 mL min^{-1} mg^{-1}, Flow STP 60 mL min^{-1}; 573 K; PCO_2 = 200 mbar
[c]TAQ500; SV 5.4 mL min^{-1} mg^{-1}, Flow STP 60 mL min^{-1}; 573 K; PCO_2 = 200 mbar

with the TAQ500 and the TGA4000 apparatus reveals that the capacities measured are very close regardless of the instrument used. Consequently, the concentration gradients between the bulk and the surface under the experimental conditions described were assumed to be negligible in both analysers. In a second set of experiments, the particle size was changed while the adsorbent mass and gas flow rate were maintained to examine the resistance to internal mass transport. Usually an increase in the CO_2 capacity with decreasing particle size reveals the presence of significant intraparticle resistance. From the results presented in Table 3.2 it is concluded that the internal transport of CO_2 is slightly restricted for particles in the size range from 250 to 500 μm and therefore it was decided to use fine powders in all subsequent tests.

The results given in Tables 3.1 and 3.2 show that after the LDH is exposed to the adsorptive gas for 1 and 2 h, the measured capacities corresponds to ~90 % and ~95 % of the value obtained after 10 h respectively. Since in multicycle experiments it is not practical to have long adsorption steps, the capacities reported in this thesis correspond to the weight gain observed either at 60 or 120 min on stream. Therefore the stability studies and the adsorption isotherms are based on working capacities which are close to the actual equilibrium values.

3.2.2 *Fixed-Bed Adsorption Column*

3.2.2.1 Experimental Setup

An experimental rig to conduct transient adsorption measurements was designed and built, Fig. 3.1. The adsorption column consisted of a stainless steel tube (d_t = 9.5 mm, ID = 7.7 mm, length = 364 mm) heated by a concentric vertical

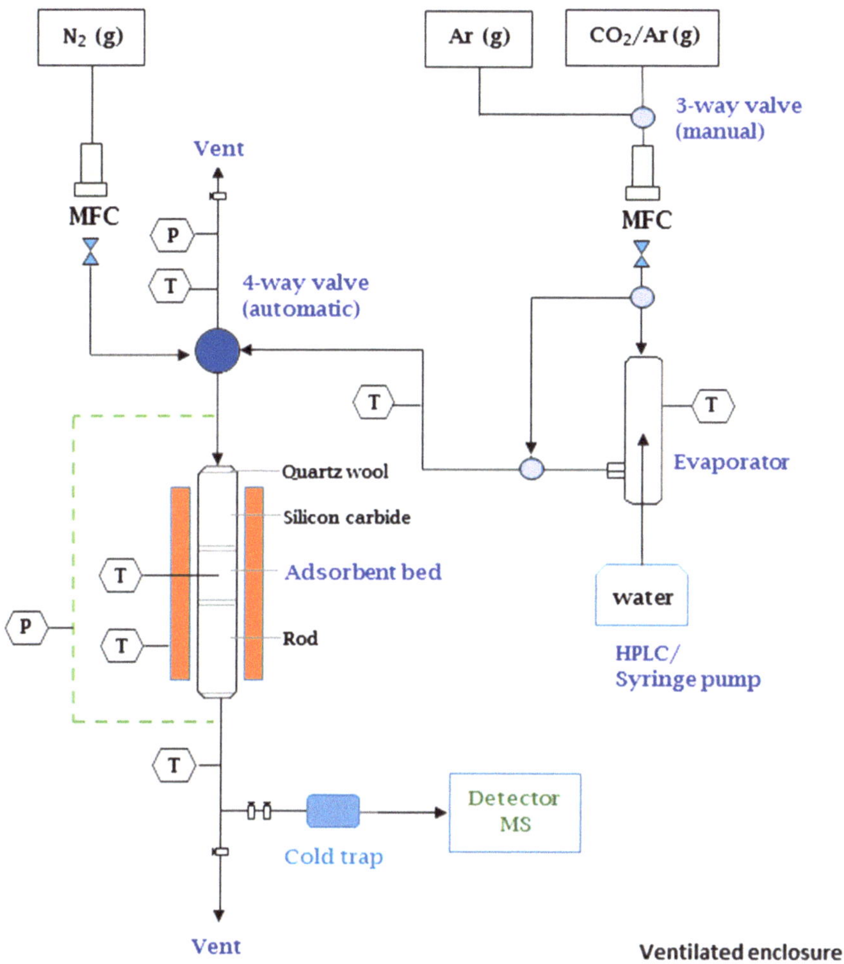

Fig. 3.1 Schematic diagram of the adsorption column system

Lenton furnace. The column was divided into 3 sections: preheat, adsorbent and exit. The preheat section was packed with silicon carbide (size 764 μm) to procure isothermal conditions and to reduce axial dispersion. In the next section 0.5–1 g of sorbent material were loaded as granules in the size range of 250–500 μm, which resulted in a bed packing around 30.2 g m^{-1}. Finally, a stainless steel hollow rod (L = 150 mm) with an orifice size of 1.6 mm was placed after the adsorbent bed, to minimize the dead volume.

The temperature of the column was measured in the centre of the adsorbent bed using a K-type thermocouple and was recorded using a Pico Technology software. Two independent feeding lines, nitrogen and CO_2/Ar mixtures (or Ar), were used to produce the desired step change by means of an electrically actuated high speed four-way switching valve (VICI). The gas flow rates were regulated by two Brooks 5850E mass flow controllers (0–0.5 L min^{-1}). Before the actuated valve, the CO_2/Ar stream could be diverted to a vaporisation system to conduct experiments in the presence of steam. Water was fed to an electrically heated column (d_t = 12.7 mm, length = 300 mm, T = 453 K) packed with silicon carbide granules (764 μm) using either an HPLC pump (Gynkotek, 0–4.99 mL min^{-1})[3] or a syringe pump (Razel, syringe 10 mL, flow range 0.397–39.3 mL h^{-1}).[4] The feeding line (stainless steel, 1.6 mm) went into the column through a bore thru T piece up to 50 mm below the top of the packed column. The CO_2/Ar mixture was fed from the top so that the dripping water had enough heat transfer surface to vaporise and mix with the flow. All the lines of the system (stainless steel, 1.6 mm) were trace heated using heating tapes to avoid condensation. The temperature of the heating tapes (usually 453 K) was regulated by an Electrothermal power regulator. The effluent of the column was monitored with a mass spectrometer (ESS GeneSys II) with heated capillary. A cold water trap with quartz beads was placed before the sampling line in tests performed under relatively high steam partial pressures.

3.2.2.2 Standard Experimental Procedure

The breakthrough tests were conducted at an absolute pressure of 1 bar. A standard adsorption experiment involved the following steps: (1) the adsorbent was activated in situ at 673 K for 4 h under N_2 (138 mL min^{-1}); (2) the N_2 flow rate and the column temperature were adjusted to the conditions of the adsorption step (typically 46–165 mL min^{-1} STP and 573 K); (3) a step change was created by switching to CO_2/Ar or CO_2/Ar/H_2O keeping the volumetric flow rate of N_2. In addition, equal pressures in the N_2 and the adsorptive gas were maintained by means of a metering valve placed in the vent of the four way valve. The adsorption step was typically carried out for 30 min under CO_2 partial pressures in the range of 17–500 mbar. (4) The adsorbent was regenerated for 30 min by isothermal N_2 purge (using the flow rate and the temperature of the adsorption step) or by temperature-swing (at 673 K purging with the same flow rate of N_2).

Adsorption kinetics and axial dispersion cause an effluent response deviation from the perfect step input. Therefore to calculate the CO_2 capacity of the adsorbents a corrective procedure was followed. The adsorption experiment was performed under the same operating conditions described above but using pure argon instead of the CO_2 containing mixture. This blank test also allowed the dead volume of the system to be determined. Other similar corrective methods have been suggested in the literature [8, 9].

[3]Typically the water flow rate was set to 0.02 mL min^{-1}.

[4]Typically the water flow rate was set to 0.397 mL h^{-1}.

3.2.2.3 Adsorption Capacity Analysis

To ensure the reliability of the results, two different methods were used to calculate the adsorption capacity of the solids. In the graphic method the evolution of the argon and CO_2 output signals with time were normalised and expressed as CO_2 mol per minute. The area between both breakthrough curves corresponded to the moles of CO_2 adsorbed and therefore were divided by the mass of adsorbent recovered from the adsorption column (m_a) to obtain the adsorption capacity of the material. Alternatively, the material balance in the column was used to calculate the moles of CO_2 adsorbed, assuming ideal gas behaviour as in 3.1 [10].

$$q_{CO_2} = \frac{Q_f P_s}{RT_s m_a}\left[t_\infty y_{CO_{2f}} - y_{Ar_f}\int_0^{t_\infty} \frac{y_{CO_2}}{y_{Ar}}dt\right] - \frac{\varepsilon_T V_b P_b y_{CO_{2f}}}{RT_b m_a} \tag{3.1}$$

In Eq. (3.1), q is the adsorption capacity of the material, Q_f is the volumetric flowrate of the feed at standard conditions, T_s and P_s are the standard temperature and pressure, $y_{CO_{2f}}$ and y_{Ar_f} are the mol fractions of CO_2 and Ar in the feed, y_{CO_2} and y_{Ar} are the mol fractions of CO_2 and Ar at the exit of the column, t_∞ is the time to approach the feed composition at the exit, and T_b and P_b are the temperature and pressure of the adsorbent bed which has total porosity ε_T and volume V_b. The derivation of 3.1 is given in Appendix B.1.

The dead-volume of the adsorption unit ($\varepsilon_T V_b$) was obtained from a blank experiment using the packed column and switching from nitrogen to argon, where

$$\varepsilon_T V_b = t_{st} Q_{Ar-bk} \tag{3.2}$$

Q_{Ar-bk} and t_{st} being the volumetric flow rate of argon and the stoichiometric time in the blank test respectively.

3.2.2.4 Mass Transfer Effects

The importance of external mass transfer limitations in an adsorption column is commonly assessed through the Biot number (Bi_m), which expresses the ratio between the internal and the external diffusional resistances in a pellet. A large Biot number, $Bi_m \gg 1$, indicates that the resistance from the bulk gas to the surface is not the controlling mass transport process (see Chap. 2, Sect. 2.2.4) [7, 11]. Biot numbers calculated for activated LDH particles under the typical experimental conditions are given in Table 3.3.

It can be observed that the fluid dynamics prevailing in the column during operation are such that the magnitude of the internal mass transfer resistances are significantly larger than the external constraints. Therefore, the relative contribution of the external mass transport to the adsorption kinetics is minimal and can be disregarded. The effect of intraparticle diffusion limitations is addressed in Chap. 7.

Table 3.3 Transport related dimensionless numbers for the adsorption column[a]

Molar composition	u (m s^{-1})	km (m s^{-1})	Sc (–)	Re (–)	Sh (–)	Bi (–)
2 % CO_2, 98 % Ar	0.03	0.27	0.99	0.30	2.53	83
2 % CO_2, 98 % Ar	0.09	0.32	0.99	0.91	3.03	100
2 % CO_2, 16 % H_2O, 82 % Ar	0.04	0.39	0.50	0.50	2.57	102
2 % CO_2, 16 % H_2O, 82 % Ar	0.11	0.47	0.51	1.48	3.11	123

[a]The Biot number was calculated as $Bi = \frac{k_m(d_{pore}/2)}{\varepsilon_p D_p + (1-\varepsilon)KD_s}$ [11]
where ε_p is the porosity of the activated LDH, d_{pore} is the pore diameter, D_p and D_s are the pore and the surface diffusion coefficients respectively, and K is the dimensionless Henry's law constant. The mass transfer coefficient, k_m, was obtained from the Sherwood number: $Sh = 2 + 1.1Re^{0.6}Sc^{1/3}$. The values of the different parameters used in the calculations are given in Appendix B.2

3.3 Catalytic Tests

3.3.1 Experimental Setup and Standard Procedure

Temperature-programmed reaction experiments were carried out to test promising water gas shift and methanol-to-shift catalysts to be used in combination with the LDH based adsorbents. The continuous tubular microreactor system depicted in Fig. 3.2 was used for that purpose. The reactor consisted of a quartz tube (d_t = 6.35 mm, wall thickness = 1.21 mm, length = 270 mm) which was heated by an

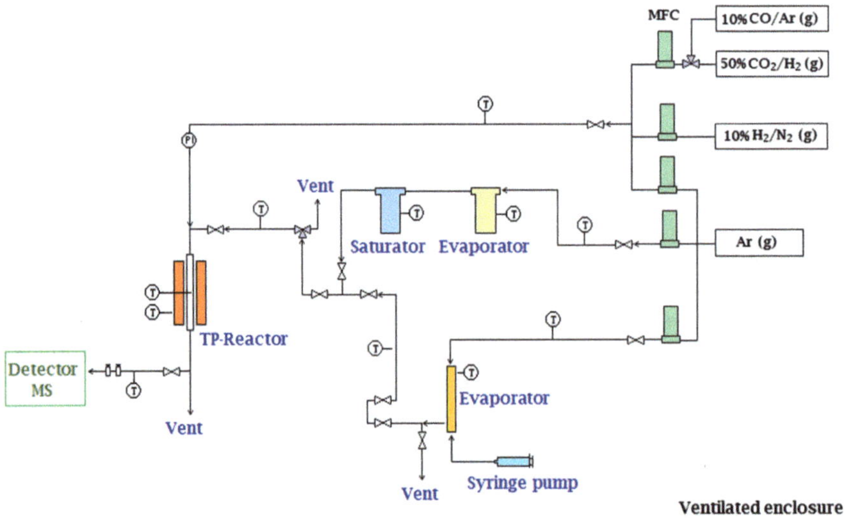

Fig. 3.2 Schematic representation of the temperature-programmed reaction system

in-house-built electric oven (d_t = 21 mm, length = 230 mm) similar to that used for temperature-programmed desorption (Sect. 3.4.7). All the lines were made of stainless steel tubing and Swagelok compression fittings were used in the connections.

The reactants were fed using mass flow controllers (Bronkhorst, EL-FLOW and Millipore, 280 SA) by two different lines, one for non-condensable gases and the other for saturated streams. In the latter case, argon was used as a carrier and was saturated either with methanol or with water. Methanol vapours were supplied by a saturator which consisted of a hot and a cold traps at atmospheric pressure filled with pure methanol and connected in series. Argon flowed through both traps forming very small bubbles to provide enough transfer area; the temperature of the cold trap set the methanol partial pressure (277 K, 50 mbar). Water was supplied using a syringe pump (Razel, syringe 10 mL, flow range 0.397–39.3 mL h^{-1}), to dose a constant amount of liquid (typically 0.397 or 1.19 mL min^{-1}) into an evaporator. The evaporator consisted of a vertical stainless steel tube (d_t = 9.5 mm, length = 145 mm) heated by electric tape and maintained at ~453 K using a thermocouple attached to the column from the outside for the control. The evaporator was filled with silicon carbide particles (740 μm). The pump was connected to the bottom of the evaporator by stainless steel tubing with a Swagelok bore thru T piece which allowed water to be fed 25 mm below the top of the packed column. Argon was fed counter currently from the top so the dripping water had enough heat-exchange area to vaporise and mix with the carrier flow. The lines connecting the saturator and the evaporator to the microreactor were trace heated at 453 K to prevent any condensation.

Commercial and in-house synthesised catalysts were tested under a temperature program for the water gas shift, reverse water gas shift and methanol decomposition reactions. Typically 30 mg of catalyst were loaded and activated for 1 h at 573 K flowing a mixture 10 % H_2 in N_2 (~50 mL min^{-1}). After the activation, the reactor was cooled down to 373 K and then the gas was changed to the desired feed composition. Subsequently, the temperature was increased with a heating rate of 10 K min^{-1} up to 973 K. For some catalysts, stability tests were carried out at 573 and 673 K for ~24 h. A small fraction of the microreactor effluent was continuously sampled and analysed by mass spectrometry (Varian Turbo V81 M). Additionally, gas samples were collected in sealed bags and analysed off-line by gas chromatography (Shimadzu GC2014) to cross-check the concentration of methane. All the experiments were performed at atmospheric pressure.

3.3.2 *Product Analysis*

3.3.2.1 Calibration of the Mass Spectrometer

The evolution of the reactions was monitored by analysing the spectra of masses produced by the species in the effluent. For the quantification of the products,

Table 3.4 Signals followed by mass spectrometry

Species	Molecule fragments (m/z)
Ar	20, 40
He	4
CO	12, 16, 28
CO_2	12, 16, 22, 28, 44
CH_4	12, 13, 14, 15, 16
H_2	1, 2
H_2O	16, 17, 18
CH_3OH	15, 31

certified gas mixtures (BOC, CP grade) were used to calibrate the signals of the instrument: mixture A (H_2 0.6 vol.%, CO 0.2 vol.%, He 1.0 vol.%, CH_4 2.0 vol.%, Ar balance); mixture B (He 2.5 vol.%, CH_4 5.0 vol.%, Ar balance); mixture C (CO_2 20.0 vol.%, Ar balance). Additionally, a stream 5 vol.% of methanol in argon (mixture D) was fed via the saturator. The fragments produced by each compound are given in Table 3.4 where the signals used to follow each species are underlined.

The mole fractions were assumed to be proportional to the intensity of the m/z signal measured and this was used to quantify the species. Additionally, to correct for slight variations in the absolute values of the MS signals, all the signals were normalised with respect to the m/z = 20 of Ar. The following expressions were obtained for carbon dioxide and methanol:

$$\frac{y_{CO_2}}{y_{Ar}} = \alpha_{CO_2/Ar} \cdot \frac{m/z = 44}{m/z = 20}. \tag{3.3}$$

$$\frac{y_{CH_3OH}}{y_{Ar}} = \alpha_{CH_3OH/Ar} \cdot \frac{m/z = 31}{m/z = 20}. \tag{3.4}$$

where $\alpha_{i/Ar}$ are the proportionality factors for the molar ratio of species i with respect to argon and m/z is the MS signal corresponding to the specified molecule fragment. For carbon monoxide and methane the MS signals coincide with fragments from other molecules or were found to be too weak to allow for an accurate quantification. Consequently, the signals m/z = 28 for CO and m/z = 15 for CH_4 had to be corrected to account for the contribution of CO_2 and methanol respectively. This can be done by recalling the fact that the ratio between the different mass fragments in a molecule is constant as long as the ionisation energy is maintained. Therefore, the proportionality factors can be obtained as follows:

$$\frac{y_{CO}}{y_{Ar}} = \alpha_{CO/Ar} \cdot \frac{\left(m/z = 28 - \frac{m/z=44}{\beta_{CO_2}}\right)}{m/z = 20} \tag{3.5}$$

$$\frac{y_{CH_4}}{y_{Ar}} = \alpha_{CH_4/Ar} \cdot \frac{\left(m/z = 15 - \frac{m/z=31}{\beta_{CH_3OH}}\right)}{m/z = 20} \tag{3.6}$$

$\alpha_{i/Ar}$ are the proportionality factors for the indicated molar ratio, β_{CO_2} is the ratio $\frac{m/z=44}{m/z=28}$ obtained from the calibration mixture C (CO_2 20.0 vol.%, Ar balance) and β_{CH_3OH} is the ratio $\frac{m/z=31}{m/z=15}$ obtained flowing the mixture D (CH_3OH 5.0 vol.%, Ar balance).

3.3.2.2 Reaction Extents

Three independent reactions were considered to occur during the temperature-programmed experiments: methanol decomposition, CO methanation and water gas shift.

$$\begin{array}{ll} CH_3OH \rightarrow CO + 2H_2 & \xi_1 \\ CO + 3H_2 \rightarrow CH_4 + H_2O & \xi_2 \\ CO + H_2O \rightarrow CO_2 + H_2 & \xi_3 \end{array}$$

The extent of reaction, ξ_j, has the units of mol per unit of time and is defined as:

$$\xi_j \equiv \frac{n_i - n_i^0}{\upsilon_{ij}}. \tag{3.7}$$

where n_i^0 and n_i are the inlet and outlet molar flowrates of species i, and υ_{ij} is the stoichiometric coefficient of i in reaction j. The ratio of the molar fraction of the different compounds to the molar fraction of argon can be written in terms of the reaction extents using the following expressions:

$$\frac{y_{CH_3OH}}{y_{Ar}} = \frac{n^0_{CH_3OH} - \xi_1}{n^0_{Ar}} \tag{3.8}$$

$$\frac{y_{CO}}{y_{Ar}} = \frac{n^0_{CO} + \xi_1 - \xi_2 - \xi_3}{n^0_{Ar}} \tag{3.9}$$

$$\frac{y_{CO_2}}{y_{Ar}} = \frac{n^0_{CO_2} + \xi_3}{n^0_{Ar}} \tag{3.10}$$

$$\frac{y_{H_2}}{y_{Ar}} = \frac{n^0_{H_2} + 2\xi_1 - 3\xi_2 + \xi_3}{n^0_{Ar}}. \tag{3.11}$$

$$\frac{y_{H_2O}}{y_{Ar}} = \frac{n^0_{H_2O} + \xi_2 - \xi_3}{n^0_{Ar}}. \tag{3.12}$$

$$\frac{y_{CH_4}}{y_{Ar}} = \frac{\xi_2}{n^0_{Ar}} \tag{3.13}$$

To calculate the reaction extents, any three of the previous equations were selected and solved simultaneously. The mass balance was checked by choosing different expressions to calculate the extents. Deviations were below 10 % and are thought to be related with small differences in the instrument sensitivity for the different mass fragments.

3.3.2.3 Conversion, Selectivity and Yield

The conversion, selectivities and yields were calculated considering the carbon containing molecules exclusively. To this end, the change in the molar flow rates of the different species throughout the experiments was obtained from the extents of reaction as follows:

$$n_{CH_3OH} = n^0_{CH_3OH} - \xi_1 \tag{3.14}$$

$$n_{CO} = n^0_{CO} + \xi_1 - \xi_2 - \xi_3 \tag{3.15}$$

$$n_{CO_2} = n^0_{CO_2} + \xi_3 \tag{3.16}$$

$$n_{H_2} = n^0_{H_2} + 2\xi_1 - 3\xi_2 + \xi_3. \tag{3.17}$$

$$n_{H_2O} = n^0_{H_2O} + \xi_2 - \xi_3 \tag{3.18}$$

$$n_{CH_4} = \xi_2. \tag{3.19}$$

The carbon conversion X is defined as[5]

$$X = \frac{n^0_{C,i} - n_{C,i}}{n^0_{C,i}} \tag{3.20}$$

where $n^0_{C,i}$ and $n_{C,i}$ are the inlet and outlet molar flowrates of the carbon containing species i included in the feed. Selectivities of the carbon products, S_i, were calculated using the following expression:

---

[5]All the carbon containing molecules in the feed contain one carbon.

$$S_i = \frac{n_i}{n^0_{C,feed} - n_{C,feed}} \tag{3.21}$$

where n_i is the outlet molar flowrate of the carbon containing product i. Finally, the carbon yield Y_i is defined as the product of X and S_i:

$$Y_i = \frac{n_i}{n^0_{C,feed}} \tag{3.22}$$

The yield of hydrogen for the methanol decomposition reaction is calculated as:

$$Y_{H_2} = \frac{y_{H_2}/y_{Ar}}{2\left(y_{CH_3OH^\circ}/y_{Ar^\circ}\right)} \tag{3.23}$$

3.4 Characterisation of Adsorbents and Catalysts

3.4.1 Nitrogen Physisorption

Physical adsorption of N_2 was carried out at 77 K in a Micrometrics Tristar 3000 instrument. The samples were dried at 393 K under N_2 for at least 12 h. The specific surface areas were obtained using the Brunauer–Emmett–Teller (BET) method and the pore volume was measured at P/Po = 0.991. The average pore diameter was calculated from the desorption isotherm using the Barrett-Joyner-Halenda (BJH) method.

3.4.2 X-Ray Powder Diffraction (XRD)

The diffraction patterns of the different materials as powder were obtained in a PANalytical X'Pert Pro multipurpose apparatus (Cu Kα radiation) operated in the reflection mode, and at 40 kV and 30 mA. The samples were placed on a silicon substrate and scanned at room temperature varying the 2-theta angles between 5° and 80°. The apparent sizes (coherence length) in the *c*-direction (reflection at $2\theta = 11.7°$) of the as-synthesised layered double hydroxides (LDHs) crystallites (supported and unsupported) were estimated using the Scherrer equation from the full width at half maximum (FWHM). After thermal treatment, the size of resulting layered double oxides (LDO) particles was calculated applying the same procedure to the diffraction peak at $2\theta = 43.2°$.

3.4.3 Transmission-Electron Microscopy/Energy Dispersive Spectroscopy (TEM/EDS)

A JEOL 2010 microscope operating at 200 kV was used to obtain high resolution transmission electron microscopy (HRTEM) images. In addition, some samples were analysed using a JEOL 2000FX instrument. The materials in fine powder were dispersed in isopropanol. A drop of the suspension was allowed to dry onto holey carbon copper or gold grids (300 mesh, Agar Scientific). Elemental analysis was done using an Oxford Instruments INCA energy dispersive X-ray spectrometer.

3.4.4 Scanning-Electron Microscopy (SEM)

The morphology of the materials was examined by scanning-electron microscopy using a Zeiss Auriga FIB/FEG-SEM microscope. Before imaging, the samples were electro-coated with gold or chromium. SEM-EDS images were obtained in a JEM-2100F instrument.

3.4.5 Inductively Coupled Plasma-Optical Emission Spectrometry (ICP-OES)

The composition of the materials was determined by measuring the ions that were stable in solution after acid digestion. The samples were ground into powder and then treated with 10 mL of *aqua regia* (concentrated HCl and HNO_3 in a ratio 3:1). After 24 h, the sample for the analysis was prepared by taking an aliquot of 5 mL from the digestion solution and diluting with deionised water to 50 mL. The detector was calibrated using standard solutions (*Trace*CERT, Fluka). The measurements were carried out in a Perkin Elmer Optima 2000DV apparatus.

3.4.6 Thermogravimetric Analysis (TGA)

The actual loadings of carbon support in the LDH hybrids (LDH/MWCNTs and LDH/GO) were measured by thermal gravimetry in a TA Instruments TAQ500 analyser. Typically, 5 mg of the sample were first dried at 393 K under nitrogen for 20 min and then heated up to 1073 K at a rate of 10 K min^{-1} flowing 20 mL min^{-1} of air. For the pure LDHs and for the carbon supports, the ratios between the mass of the dry sample and the remaining mass at the end of the thermal analysis (hereafter called residues) were found to be reproducible and characteristic of each material (i.e. $\sim$0.6 for the LDH and $\sim$0.1 for the carbon supports). Consequently, it was possible to determine the actual loading of carbon support by assuming that the LDH and support residues were additive and equal to the residue of the hybrid. The masses of LDH and support were considered to be the only contributors to the mass of the dry hybrids. This can be expressed as follows:

$$x + y = 1 \tag{3.24}$$

$$x \cdot r_{LDH} + y \cdot r_{support} = r_{hybrid} \tag{3.25}$$

where x and y are the mass fractions of LDH and support respectively, and r_{LDH}, $r_{support}$ and r_{hybrid} are the residues of the LDH, the support and the hybrid.

3.4.7 Temperature-Programmed Desorption (TPD)

As part of the present project, a rig was designed and commissioned to carry out CO_2-TPD experiments. Argon (BOC, CP grade) was used in the sample pre-treatment and during desorption while a premixed certified gas 20 % CO_2 in Ar (BOC, CP grade) was used for the adsorption. Before entering the reactor, the argon passed through oxygen (Oxisorb, Messer Griesheim) and water (Hydrosorb, Messer Griesheim) traps for further purification. The gas flow rates were adjusted with manual flow meters. Stainless steel tubing was used for the lines and Swagelok compression fittings for the connections. The sample was loaded into a quartz tube microreactor (d_t = 6.35 mm, wall thickness = 1.21 mm, length = 270 mm) which was heated by an in-house-built electric oven (d_t = 21 mm, length = 230 mm) consisting of a wire coiled around a quartz cylinder and wrapped with various layers of quartz wool. The heating rate in the oven was controlled using a thermocouple attached to the inner wall of a quartz cylinder. The temperature was recorded using a second thermocouple attached to the microreactor from the outside at the height where the adsorbent bed was placed. The temperature profile of the oven was evaluated so that the adsorbent bed was positioned in the isothermal region. The effluent of the reactor was analysed online by mass spectrometry (MS) with an ESS GeneSys II instrument with a heated capillary.

To carry out the experiments, 20–30 mg of the sample were preconditioned by heating from room temperature to 673 K at 10 K min^{-1} under a flow of argon (45 mL min^{-1}). After cooling to 313 K, the sample was exposed to the CO_2/Ar mixture for 1 h. Subsequently, the reactor was purged with argon for 2 h to remove weakly adsorbed CO_2 and then the temperature was increased to 1023 K at 50 K min^{-1}. CO_2 desorption was followed by MS with the signal m/z 44. Additionally, the signals for water (m/z 18), CO (m/z 28), argon (m/z 40, 20) and oxygen (m/z 32) were also monitored. Prior to each TPD experiment, the MS response was calibrated by injecting the premixed CO_2 gas using an electrically actuated 6 port valve (VICI) and a certified 2 mL loop.

3.4.8 Temperature-Programmed Reduction (TPR)

To determine an adequate pretreatment protocol for the activation of the water gas shift and methanol decomposition catalysts, reduction was conducted under a temperature program in the quartz microreactor system previously described for TPD. The catalyst sample ($\sim$50 mg) was loaded in the quartz tube as a powder and purged with argon at 300 K for 20 min. Next, the feed gas was switched to a mixture 10 % H_2/N_2 (45 mL min^{-1}) and the temperature increased to 823 K at a heating rate of 10 K min^{-1}. The hydrogen consumption was monitored with a thermal conductivity detector (TCD) calibrated using pure CuO (purity 99.9999 %, Alfa Aesar).

3.4.9 Particle Size Distribution, Helium Pycnometry and Mercury Porosimetry Techniques

The particle size distribution and the mean particle size of some LDH-based adsorbents were determined using a Malvern analyser (Mastersizer 2000). Aqueous solutions containing small amounts of surfactant (dish soap) were used as liquid dispersing medium. The skeletal density of the materials was determined by helium pycnometry using an AccuPyc 1330 instrument. In addition, some other properties of the adsorbents such as bulk density, porosity and tortuosity were measured by mercury intrusion porosimetry (Micromeritics AutoPore IV). In all the techniques at least 250 mg of sample, either powder or pellets, were used to carry out the corresponding tests.

References

1. Suzuki, M. (1990). *Adsorption engineering*. Tokyo, Japan: Kodansha.
2. Hutson, N. D., & Attwood, B. C. (2008). High temperature adsorption of CO_2 on various hydrotalcite-like compounds. *Adsorption, 14*(6), 781–789.
3. Du, H., Ebner, A. D., & Ritter, J. A. (2010). Temperature dependence of the nonequilibrium kinetic model that describes the adsorption and desorption behavior of CO_2 in K-promoted HTlc. *Industrial and Engineering Chemistry Research, 49*(7), 3328–3336.
4. Ruthven, D. M. (1984). *Principles of adsorption and adsorption processes*. NY, USA: John Wiley and Sons.
5. Wood, G. O. (2002). Quantification and application of skew of breakthrough curves for gases and vapors eluting from activated carbon beds. *Carbon, 40*(11), 1883–1890.
6. Ding, Y., & Alpay, E. (2000). Equilibria and kinetics of CO_2 adsorption on hydrotalcite adsorbent. *Chemical Engineering Science, 55*(17), 3461–3474.
7. Yang, R. T. (1997). *Gas separation by adsorption processes* (Vol. 1). London, UK: Imperial College Press.
8. Choi, S., Drese, J. H., & Jones, C. W. (2009). Adsorbent materials for carbon dioxide capture from large anthropogenic point sources. *ChemSusChem, 2*(9), 796–854.

9. Lively, R. P., Chance, R. R., Kelley, B. T., Deckman, H. W., Drese, J. H., Jones, C. W., et al. (2009). Hollow fiber adsorbents for CO_2 removal from flue gas. *Industrial and Engineering Chemistry Research, 48*(15), 7314–7324.
10. Lee, K. B., Verdooren, A., Caram, H. S., & Sircar, S. (2007). Chemisorption of carbon dioxide on potassium-carbonate-promoted hydrotalcite. *Journal of Colloid and Interface Science, 308* (1), 30–39.
11. Do, D. D. (1998). *Adsorption analysis: Equilibria and Kinetics*. London, UK: Imperial College Press.

Chapter 4
Layered Double Hydroxides Supported on Multi-walled Carbon Nanotubes for CO_2 Adsorption

4.1 Introduction

The development of materials with adequate CO_2 adsorption characteristics for the sorption enhanced water gas shift reaction remains challenging as discussed in Chap. 2. In this regard layered double hydroxides (LDHs)[1] are the most promising group of CO_2 adsorbents showing high selectivity towards CO_2 at temperatures between 473 and 723 K. Additionally, in this operating window LDHs exhibit fast adsorption-desorption kinetics, good performance in the presence of water and, compared to other potential chemisorbents (e.g. Li_2ZrO_3 and CaO based materials), they can be regenerated under milder conditions. However, despite these positive properties, there are still some major drawbacks to overcome before LDH adsorbents can be commercialised. Mainly, efficient methods to increase their relatively low adsorption capacity and to improve their multicycle stability need to be found [1, 2].

Recently, it has been observed that the CO_2 adsorption performance of LDHs can be considerably enhanced by supporting/combining them with high surface area materials such as zeolites [3], alumina [4] and carbon nanofibers (CNFs) [5]. In particular the use of CNFs has been reported to increase significantly the adsorption capacity per mass of activated Mg-Al LDHs at 523 K [5, 6]. The experiments were conducted over hybrids containing a large fraction of CNFs (82–95 wt%) and the positive effect was attributed to the formation of $Mg(Al)O_x$ nanoparticles with a high number of low-coordination sites located at the edges and corners, which have stronger basicity compared to the basal planes. No loss of capacity was observed in

[1] After thermal treatment layered double hydroxides (LDHs) are transformed into layered double oxides (LDOs), which are active for CO_2 adsorption at relatively high temperature (see Chap. 2, Sect. 2.4.1).

D. Iruretagoyena Ferrer, *Supported Layered Double Hydroxides as CO_2 Adsorbents for Sorption-enhanced H_2 Production*, Springer Theses,
DOI 10.1007/978-3-319-41276-4_4

multiple adsorption-desorption cycles in the presence of water. While these results encourage the use of carbon nanostructures as supports for layered double hydroxides (LDHs), structurally carbon nanofibres are not the most obvious choice due to their tendency to fracture. Additionally, the edge endings of the CNFs provide sites for oxidation and chemical functionalisation which may cause a gradual degradation of the support [7, 8].

The present chapter addresses the use of multi-walled carbon nanotubes (MWCNTs) as support for LDH derivatives with application in CO_2 adsorption. Compared to CNFs, the morphology of carbon nanotubes is better defined, typically with smaller diameter and hence a higher accessible surface to act as support. Furthermore, the aspect of MWCNTs is not affected during processing, handling and usage, maintaining an effective open framework [8, 9]. The LDH loadings used here are significantly higher than those reported in previous works [4–6, 10]. Considering the cost of the supports and the size of the adsorption units, high loadings of LDH are more relevant for industrial applications. Based on the information available in the literature this is the first time that LDH/MWCNTs are studied for CO_2 adsorption. However, there have been few studies regarding the synthesis of these types of composite-materials for their potential use in the reinforcement of polymers and as flame retardants for polypropylene [11, 12].

This chapter begins by explaining the methodology used to synthesise Mg-Al LDHs and LDH/MWCNTs hybrids (Sect. 4.2). In addition the procedure followed to activate the adsorbents prior to the CO_2 adsorption measurements is described. Then the structural and physical properties of the adsorbents are studied by a range of characterisation techniques including TGA, TEM, SEM, XRD, BET and CO_2-TPD (Sect. 4.3). The influence of the MWCNTs on the CO_2 adsorption capacity of the synthesised hybrids under dry conditions and their performance after continuous adsorption-desorption cycling is reported in Sect. 4.4. Finally, a brief summary and critique of the findings discussed along the chapter is given in Sect. 4.5. Additionally, areas for further research are identified.

The studies included in this chapter were done in collaboration with the Materials Chemistry group of Professor Milo Shaffer from the Department of Chemistry at Imperial College London. The procedures to prepare the materials were developed by the Chemistry group. In particular, the synthesis of the carbon supported materials in small scale[2] (and the titration of carboxylic acid groups on the MWCNTs) was conducted by Dr. Ainara Garcia-Gallastegui.

[2]All the adsorbents presented in this chapter were prepared as part of this PhD project in the Department of Chemistry following the procedures developed by the Shaffer's group. The carbon supported hybrids in small scale ($\sim$0.8 g) were synthesised by Dr. Garcia-Gallastegui.

4.2 Synthesis of the Adsorbents

4.2.1 Unsupported Mg-Al LDHs

The unsupported layered double hydroxides were synthesised by co-precipitation under high supersaturation conditions (see Chap. 2, Sect. 2.4) with a molar Mg/Al ratio of 2, which has been reported to be adequate for CO_2 adsorption [13]. To prepare 1 g of material, 4.5 mL of an aqueous solution containing 9 mmol Mg $(NO_3)_2{\cdot}6H_2O$ and 4.5 mmol $Al(NO_3)_3{\cdot}9H_2O$ was added dropwise to an aqueous solution (6.7 mL) containing 8 mmol Na_2CO_3 and 32 mmol NaOH under vigorous stirring (300 rpm). The salts were purchased from Sigma-Aldrich. During the addition, the temperature of the solution was maintained at 333 K and the final pH obtained was 10. The resulting white suspension was aged for 12 h keeping the stirrer speed and the temperature of the addition. The precipitate was filtered using polycarbonate membranes of 0.4 μm (Milipore) and washed with 500 mL of water at 333 K. The samples were dried overnight at 393 K under vacuum. Typically 0. 5–3 g of unsupported LDHs were synthesised per batch. Similar procedures to synthesise LDHs have been reported in the literature [5, 14].

4.2.2 Nanostructured LDH/MWCNT Hybrids

The as received MWCNTs are likely to contain amorphous carbon, graphitic nanoparticles and residues of the metals that were used to catalyse their growth by vapour deposition [8, 9]. To remove these impurities, the nanotubes are usually treated with acids or other oxidising agents and then washed with dilute bases to eliminate the oxidative debris left by the acid treatment [15]. This method introduces oxygen-containing acid groups which are converted into their conjugate salts when the base is added. These carboxylate anions facilitate the aqueous dispersion of the MWCNTs under alkaline conditions and are expected to favour the deposition of LDH nanocrystals during the coprecipitation of Mg^{2+} and Al^{3+} ions.

The present MWCNTs (CVD-grown with average diameter of 10–15 nm, ARKEMA) were oxidised using 7 mL of a mixture of concentrated H_2SO_4 and HNO_3 in a volume ratio 3–1 for every 200 mg of carbon treated. The mixture was stirred and refluxed for 30 min at 393 K. The suspension was then filtered using 0.4 μm polycarbonate membranes and washed with 500 mL of NaOH (0.01 M). The nanotubes were washed with distilled water until a neutral pH was obtained (Fig. 4.1). The oxidised MWCNTs were dispersed into 2.06 mL of an aqueous solution containing 9.9 mmol NaOH and 2.5 mmol Na_2CO_3. The amount of carbon nanotubes added depended on the desired loading of MWCNTs in the hybrid. Subsequently, 1.4 mL of an aqueous solution containing 2.8 mmol Mg $(NO_3)_2{\cdot}6H_2O$ and 1.4 mmol $Al(NO_3)_3{\cdot}9H_2O$ were added dropwise. The resulting black suspension was aged, filtered, washed and dried as described earlier for the

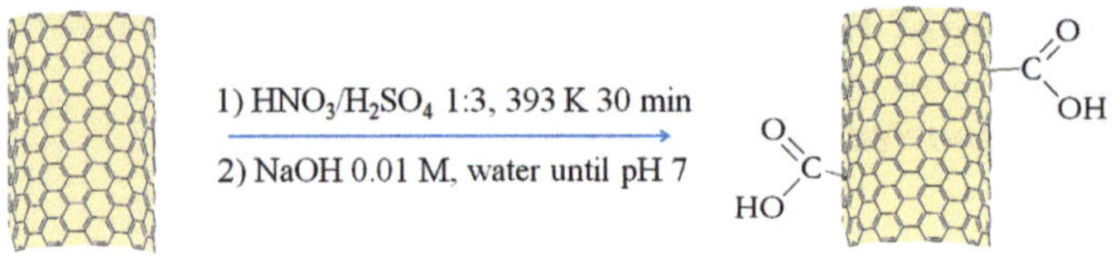

Fig. 4.1 Procedure followed to oxidise the MWCNTs

unsupported LDH (Sect. 4.2.1). Four different LDH/MWCNT hybrids were prepared with nominal weight ratios 4/1 (LDH_4), 2/1 (LDH_2), 1/1 (LDH_1) and 0.5/1 ($LDH_{0.5}$), using, 77.5, 155, 310 and 620 mg of oxidised MWCNTs respectively at constant LDH precursor concentration. To check for the consistency of the properties of the hybrids produced in larger scale, a batch of 3 g of LDH_1 was prepared by scaling all the values and then it was characterised.

4.2.3 Activation of Pure LDH and LDH/MWCNT Hybrids

Crystalline LDHs show low CO_2 adsorption capacities and therefore they need to be activated by a thermal treatment to form pseudo-amorphous layer double oxides (LDOs).[3] The unsupported and supported as synthesised materials were calcined ex situ at 673 K for 4 h flowing 100 mL min^{-1} of N_2 using a tubular quartz reactor (ID = 5 cm) placed in a horizontal concentric Carbolite furnace. The corresponding activated adsorbents are denoted as LDO, LDO_4, LDO_2, LDO_1 and $LDO_{0.5}$. The calcination temperature was chosen based on studies by Reddy et al. [16] and Hutson et al. [17] who reported that LDH derivatives produced at 673 K show an optimum balance between surface area and basic sites which maximises their CO_2 adsorption capacity and favours reversible adsorption.

4.2.4 Titration of the Oxidised MWCNTs[4]

As it was mentioned previously, the ionic interactions between the negatively-charged oxidised nanotubes and the Mg^{2+} and Al^{3+} cations in solution are expected to be crucial in the deposition and LDH growth during the coprecipitation. To estimate the extent of such interactions, the concentration of carboxylic acid groups introduced onto the MWCNTs by the acid treatment was

[3]Readers are referred to Chap. 2, Sect. 2.4.1, for further details.

[4]The titration of the oxidised MWCNTs was carried out by Dr. Garcia-Gallastegui and is included for the sake of completeness of the chapter.

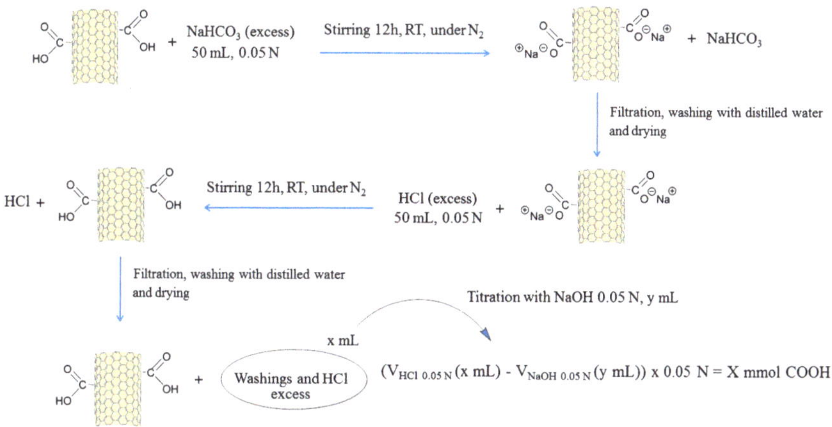

Fig. 4.2 Determination of carboxylic acid groups on the oxidised MWCNTs by reverse titration

determined by the Boehm titration method [18, 19]. A schematic representation of the procedure is presented in Fig. 4.2.

The concentration of carboxylic acid groups on the oxidised carbon nanotubes was found to be 3.4 mmol g^{-1}, which corresponds to a surface charge density of 18 (μmol e^{-}) m^{-2}.[5] For the LDH, the extra positive charge due to the substitution of Al^{3+} is equivalent to 0.33 electrons per mol and is contained within the area of the octahedral unit, i.e. 8.067 Å^2 [20]. Therefore, the excess surface charge density of LDH is approximately 6.7 (μmol e^{+}) m^{-2}, which is on the same order of magnitude that the surface charge density of the nanotubes. Hence, the idea that LDH tends to nucleate on and remain associated with the MWCNT surface is plausible.

4.3 Characterisation

The unsupported and supported LDHs in the as synthesised and activated forms were characterised using a range of physicochemical techniques. In particular the samples were analysed by X-ray diffraction (XRD), transmission electron microscopy (TEM), scanning electron microscopy/energy dispersive spectrometry (SEM/EDS), thermogravimetric analysis (TGA), N_2 physisorption at 77 K and temperature-programmed desorption of CO_2 (CO_2-TPD). The detailed description of the equipment and the test procedures was given in Chap. 3 (Sect. 3.4).

[5]This value was obtained using the surface area of the oxidised MWCNTs (191 m^2 g^{-1}).

4.3.1 Composition of the Adsorbents

The composition of the unsupported LDH and the LDH/MWCNT hybrids was determined by in situ EDS and thermogravimetric analysis. Energy dispersive spectroscopy confirmed the presence of magnesium and aluminium close to the intended molar ratio of two. The actual values measured for the different samples prepared are given in Table 4.1.

The weight losses under flowing air of the pure LDH, the MWCNTs and the hybrids in the temperature range from 373 to 1073 K are presented in Fig. 4.3. The pure LDH exhibits three stages of weight loss. The first stage, around 483 K, is attributed to the removal of weakly-held water molecules from the LDH interlayer and is followed by a weight loss between 483 and 733 K due to the removal of hydroxyl and carbonate groups present in the interlayer space. The very gradual decay thereafter may be related to some decarbonation of the remaining carbonate anions. The oxidation of the carbon nanotubes, on the other hand, occurs in single stage in a wide range of temperatures, from 500 to 800 K. The thermograms of the hybrids show a combination of features of the LDH and the support.

The actual content of CNT support in the hybrids can be estimated from the remaining mass at the end of the thermogravimetric analysis assuming that the residues of each phase are additive (see Chap. 3, Sect. 3.4.6). For the LDH, the residual mass was found to be around 60 wt% and consists of a mixture of solid magnesium and aluminium oxides. The residue of the MWCNTs accounts for 8 wt% of the initial mass loaded and is composed by metallic nanoparticles that were originally used as catalyst for the nanotubes growth. The measured MWCNTs weight contents of the hybrids are included in Table 4.1 and are systematically lower than the nominal support loadings. This is probably due to loss of carbon nanotubes during filtration. It is noteworthy that the scale up of the preparation of the LDH_1 hybrid did not introduce additional discrepancies with respect to the intended nominal composition. By combining the TGA and the EDS results, the molecular formula of the LDH dehydrated sample is found to be $Mg_{0.66}Al_{0.33}(OH)_2(CO_3)_{0.15}{\cdot}nH_2O$; where the water content estimated from the TGA corresponds to $n = 0.5$ mol (see Appendix C.1 for further details).

Table 4.1 Composition of the as-synthesised adsorbents

Sample name[a]	wt% MWCNTs nom.	wt% MWCNTs actual	Mg/Al (mol/mol)
Pure LDH	0	0	2.2
LDH_4	20	18	2.2
LDH_2	33	27	2.1
LDH_1	50	38	2.2
LDH_1 (larger scale)	50	39	1.9
$LDH_{0.5}$	67	67	2.2

[a]LDH_x where x is the nominal LDH to MWCNTs weight ratio of the sample

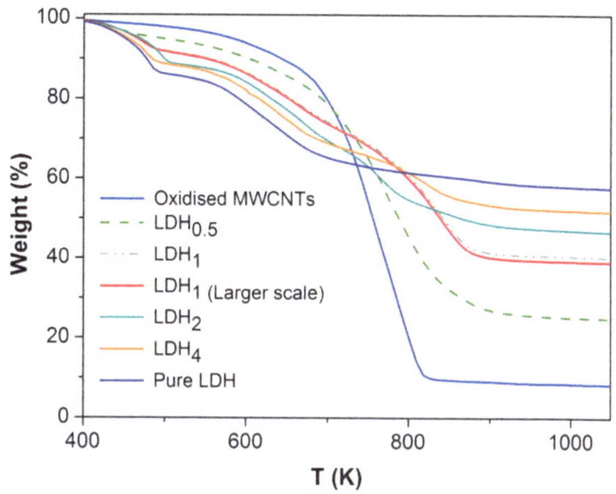

Fig. 4.3 Thermograms of pure LDH, oxidised MWCNTs and LDH/MWCNT hybrid samples

4.3.2 *Crystal Morphology*

The X-ray diffraction patterns of the oxidised MWCNTs, the unsupported LDH and the hybrids are presented in Fig. 4.4. The LDH-containing samples show the characteristic reflections of two-dimensional hydrotalcite-like compounds and were indexed accordingly (JCPDS No. 14-191) [21]. The oxidised nanotubes and the hybrids exhibit a diffraction peak at $2\theta = 26.3°$ which corresponds to the (002) plane of graphitic carbon (JCPDS No. 41-1487).

The XRD results in Fig. 4.4 indicate that no crystalline impurities are present in the materials and that the hybrids contain the intended phases. The intensity of the LDH reflections relative to the MWCNT decreases with the increasing loading of carbon support as expected. Additionally, the (003) LDH peak, which corresponds

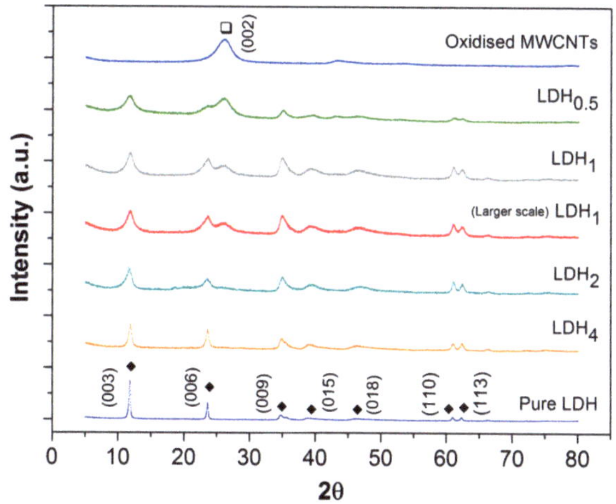

Fig. 4.4 Representative XRD diffraction patterns of as-synthesised adsorbents. Characteristic reflections of LDH (♦) and MWCNTs (❑)

Table 4.2 Crystallite size of the as-synthesised adsorbents

Sample name[a]	Crystallite size (003) (nm)
Pure LDH	30
LDH_4	17
LDH_2	16
LDH_1	7
LDH_1 (larger scale)	9
$LDH_{0.5}$	6

[a] LDH_x where x is the nominal LDH to MWCNTs weight ratio of the sample

to the layer-stacking direction (*c*-direction), broadens with increasing MWCNT contents. The apparent crystallite sizes obtained by the Scherrer equation from the (003) reflections are given in Table 4.2. The significant decrease in the calculated size between LDH_4 (17 nm) and $LDH_{0.5}$ (6 nm) can be due to differences in the thickness of the LDH platelets or can result from structural disorder in that specific direction. LDH_1 prepared in larger scale falls into this trend. The peak positions in the hybrid materials do not change with respect to the pure LDH. Therefore, variations in the interlayer spacing due to intercalation of CNTs or to hydroxyl-mediated interactions between LDH and the support are not relevant [22, 23].

After activation of the pure LDH and the hybrids, the double layered crystalline structure collapses, Fig. 4.5. Only weak broad peaks at 37°, 43° and 62° are observed which correspond to diffractions by the (111), (200) and (220) planes of MgO periclase (JCPDS No 45-946). Aluminium oxide is assumed to be well-dispersed in the oxide mixture [24].

Crystalline hexagonal platelets, typical of layered double hydroxides, are observed in the TEM images with an average lateral size and crystal thickness about 30 and 10 nm respectively, Fig. 4.6. The diameter of the nanotubes was around

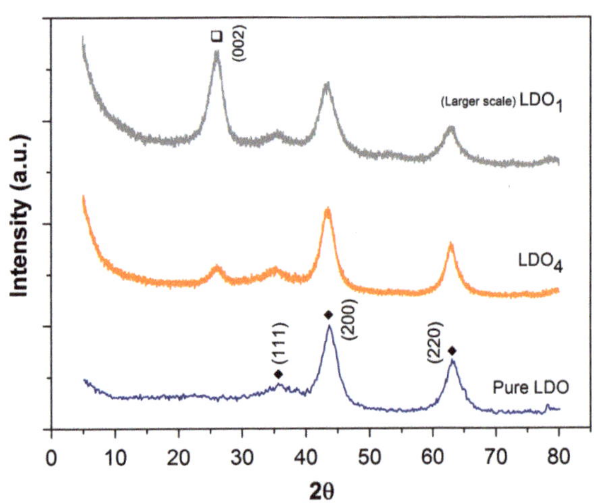

Fig. 4.5 Representative XRD diffraction patterns of activated adsorbents. Characteristic reflections of LDO (♦) and calcined MWCNTs (❑)

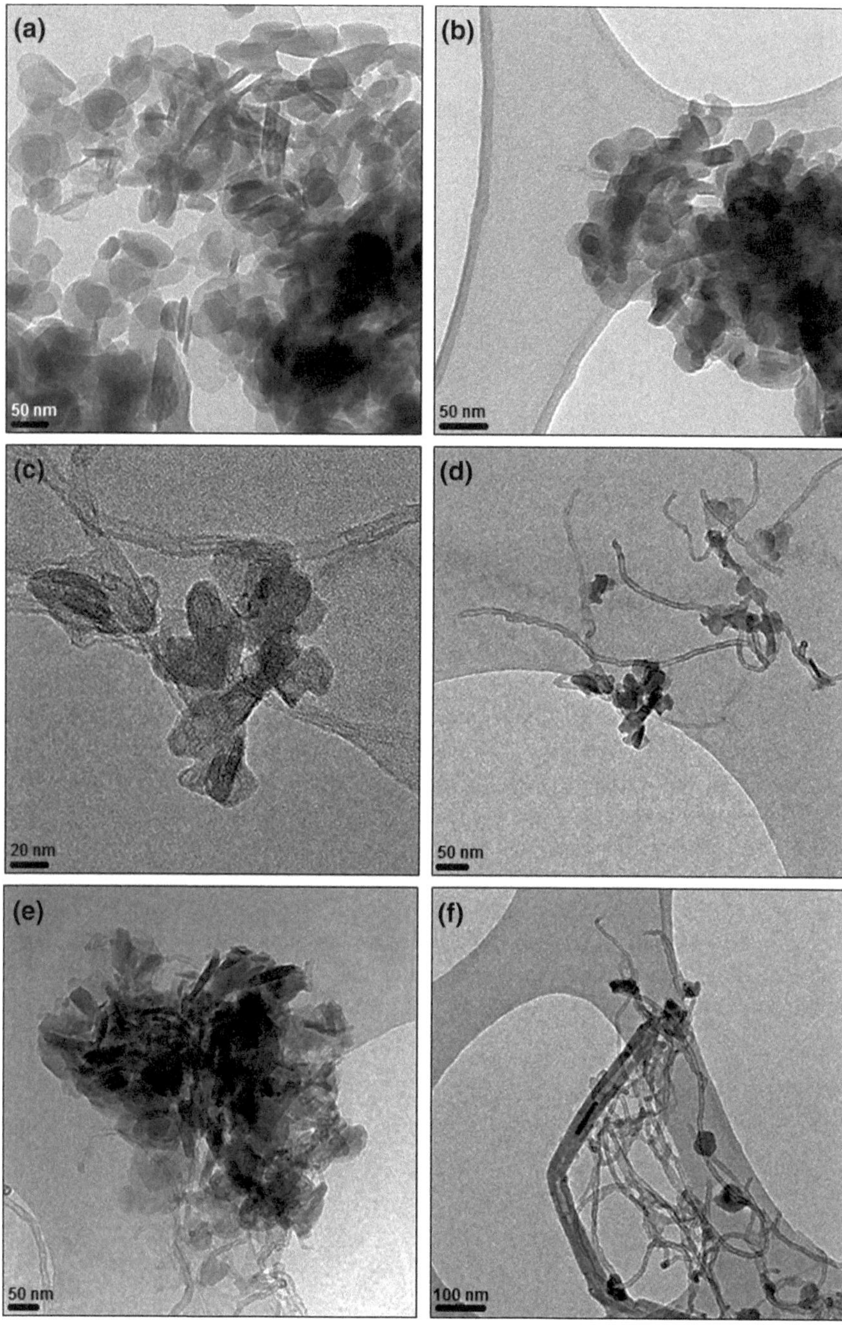

Fig. 4.6 Representative TEM images of pure LDH (**a**), LDH_4 (**b**), LDH_2 (**c** and **d**), LDH_1 (**e**) and $LDH_{0.5}$ (**f**)

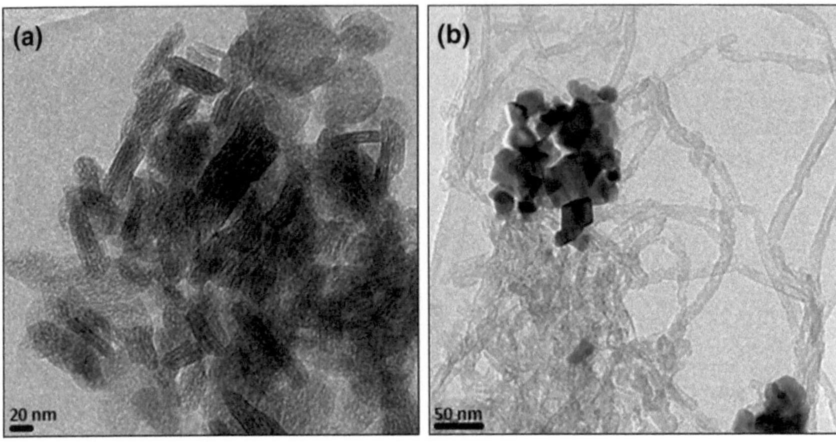

Fig. 4.7 Representative TEM images of pure LDO (**a**) and $LDO_{0.5}$ (**b**)

10 nm. In all the micrographs obtained, the LDH particles are associated (alone or in clusters) with the CNT structures and no independent agglomerates were observed. This confirms the existence of a good charge interaction between the nanotube surface and the metal cations during the coprecipitation as discussed in Sect. 4.2.2. Since the TEM images do not show large variations in the thickness of the LDH crystallites for the different hybrids, it is assumed that the appearance of stacking faults with increasing CNT loadings contributes significantly to the peak broadening of the XRD patterns. TEM images of the pure LDO and the $LDO_{0.5}$ hybrid did not reveal any significant modification in the morphology of the adsorbents after thermal treatment, Fig. 4.7.

SEM images of the dry LDH/MWCNT hybrids show a porous, agglomerated powder with a uniform dispersion of MWCNT through the sample, Fig. 4.8. Increasing the loading of LDH in the hybrids results in an increase in the number of particles observed but no other morphological differences were evidenced. In agreement with TEM, SEM images showed no marked differences in morphology

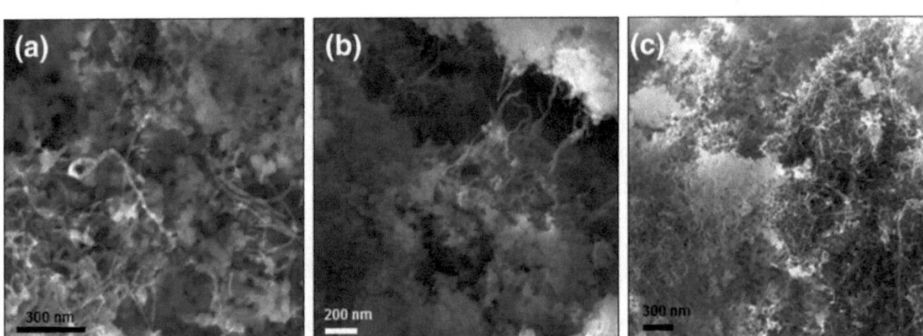

Fig. 4.8 Representative SEM images of LDH_2 (**a**), LDH_4 (**b**) and LDO_4 (**c**)

between the as-synthesised and activated materials. Similar observations can be found in the literature [25].

4.3.3 *Textural Properties*

The N_2-sorption isotherms of the oxidised CNTs and the as-synthesised pure LDH and hybrids are given in Fig. 4.9. The N_2 uptake occurs mainly in the mesoporous region of the curve; the micropore volume is negligible, in agreement with previous

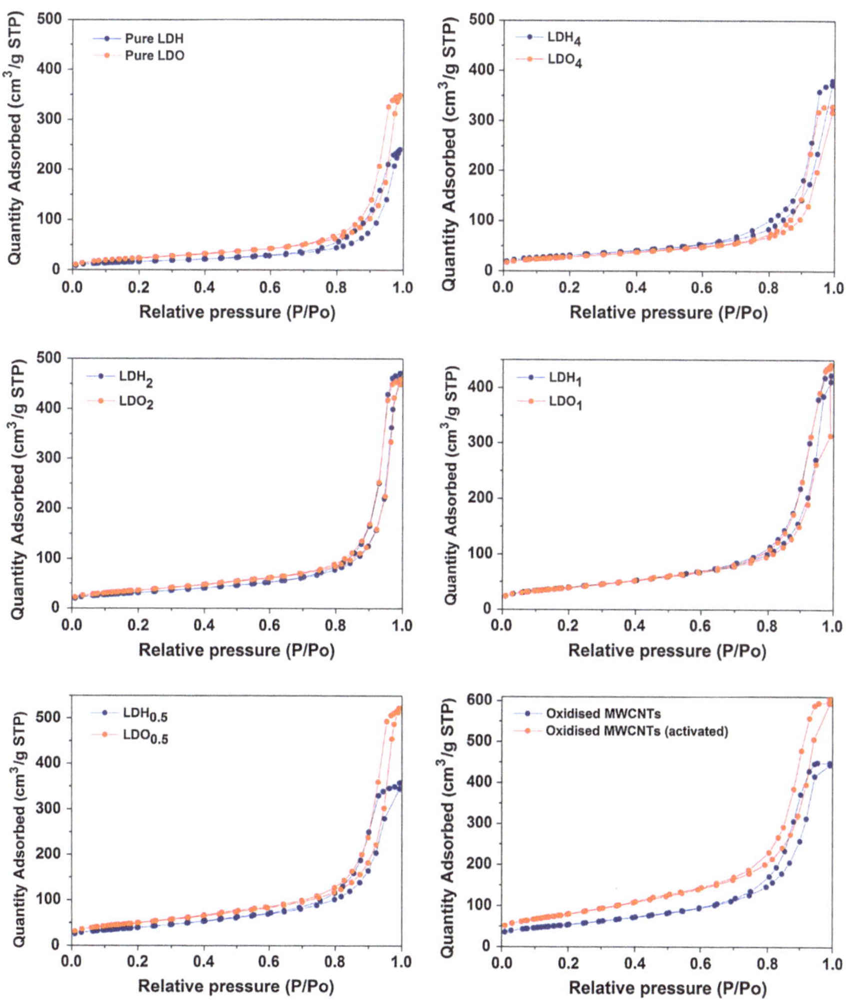

Fig. 4.9 N_2 physisorption isotherms of as-synthesised and activated adsorbents

reports [5]. The isotherms present a type H3 hysteresis loop which is usually given by aggregates of platy particles [26]. Therefore, the majority of the pore volume is attributed to interparticle voidage although a small contribution from the hollow core of the MWCNTs or the interlayer LDH galleries cannot be dismissed.

The BET surface areas and the pore diameters and volumes obtained by the BJH method are given in Table 4.3. It can be observed that the pore size distribution shifts towards larger values as the LDH content in the hybrids increases although the mesopore volume remains almost constant. The BET surface area of the pure LDH was found to be 64 $m^2\ g^{-1}$ and increases with increasing nanotube content. While it is difficult to determine whether the surface area follows a simple rule of mixtures or whether there is a subtle systematic trend due to changes in intrinsic crystal morphology, it is clear that the pronounced changes in the LDH surface area that would accompany a decrease in crystal thickness from 17 to 6 nm are not present. The activation of the LDH-containing samples do not alter the specific surface area significantly but the calcination of the oxidised MWCNTs result in a marked increase which is attributed to the removal of surface oxides thereby facilitating the access of nitrogen, Table 4.3. The N_2 physisorption isotherms of the activated adsorbents are presented in Fig. 4.9. The isotherms of the hybrid prepared in larger scale (LDH_l and LDO_l) are shown in Appendix C.2.

4.3.4 *Temperature-Programmed Desorption of CO_2*

The TPD profile of CO_2 adsorbed on the unsupported LDH after activation at 673 K is displayed in Fig. 4.10. A blank run over the activated sample is also presented.[6] The CO_2 uptake at 313 K of the calcined LDH (calculated subtracting the corresponding blank) is 0.44 mol $CO_2\ kg^{-1}$ LDO.

The evolution of CO_2 occurs in overlapping peaks with the maxima centred at 423 K (α), 543 K (β), and a high temperature stage that can be deconvoluted in two Gaussian peaks at 813 K (γ) and 963 K (δ), Fig. 4.11.[7] The strength of the basic sites in the surface increases with the CO_2 desorption temperature. The low temperature peak has been previously assigned to the release of CO_2 from bicarbonates formed on Brønsted OH^- groups and is followed by the contribution of bidentate carbonates bonded to metal-oxygen pairs (intermediate temperature). The high temperature states correspond to monodentate carbonates adsorbed on low-coordination oxygen anions [27, 28]. In the present TPD experiment, the low and intermediate energy states are the main contributors to the total adsorption capacity of the activated LDH. Feasible uses of activated LDHs can be found at

[6] In the blank experiment the unsupported LDO was not exposed to CO_2 adsorption. The pre-calcined sample was pretreated at 673 K during 1 h under flowing argon. Then the temperature was decreased to 313 K and the CO_2 evolved during a temperature program was monitored by MS.

[7] The deconvolution of the TPD profile was carried out using Fityk 0.8.9.

Table 4.3 Textural properties of as-synthesised and activated adsorbents

Sample name[a]	As-synthesised adsorbents			Sample name[a]	Activated adsorbents		
	SBET ($m^2\ g^{-1}$)	Vpore ($cm^3\ g^{-1}$)	dpore (nm)		SBET ($m^2\ g^{-1}$)	Vpore ($cm^3\ g^{-1}$)	dpore (nm)
Pure LDH	64	0.32	30	Pure LDO	87	0.47	20
LDH_4	112	0.57	17	LDO_4	101	0.49	17
LDH_2	111	0.56	22	LDO_2	128	0.52	19
LDH_1	142	0.59	16	LDO_1	139	0.48	16
LDH_1 (larger scale)	159	0.65	16	LDO_1 (larger scale)	148	0.58	15
$LDH_{0.5}$	161	0.62	11	$LDO_{0.5}$	178	0.53	13
Oxidised MWCNTs	191	0.68	12	Oxidised MWCNTs	285	0.92	11
Commercial MWCNTs	223	0.76	24	Commercial MWCNTs	–	–	–

[a]LDH_x and LDO_x where x is the nominal LDH to MWCNTs weight ratio of the sample

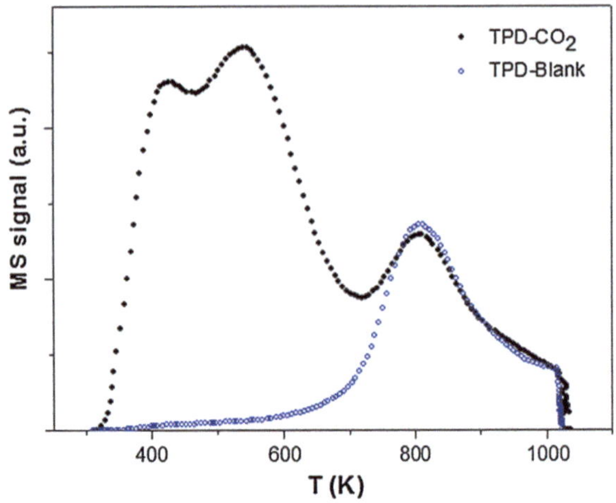

Fig. 4.10 CO_2-TPD of the activated LDH following pretreatment at 673 K and adsorption at 313 K. The blank was subjected to pretreatment only

temperatures near 423 K for low temperature applications (e.g. CO_2 recovery from flue gases) and near 543 K for processes requiring higher operating temperatures (e.g. sorption-enhanced hydrogen production). In the former case, full advantage of the basicity is taken, whereas the latter case benefits from a low regeneration temperature gradient (543–673 K) which is desirable for temperature-swing

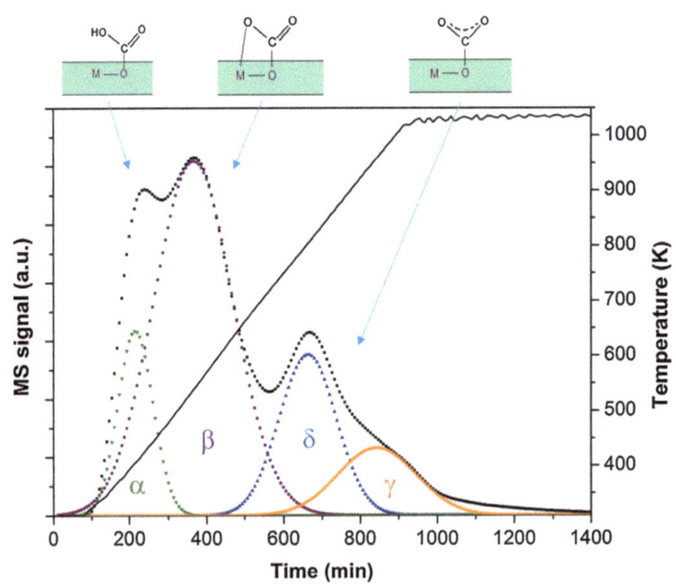

Fig. 4.11 Deconvolution of the CO_2-TPD of pure LDO

processes. This study focuses on intermediate temperature applications and therefore adsorption and desorption temperatures of 573 and 673 K respectively were considered to assess the performance of the adsorbents. Under these conditions, it is mainly the desorption state centred at 543 K that is cyclically populated and depopulated.[8]

4.4 CO_2 Adsorption Experiments

The CO_2 adsorption capacity and thermal stability of the adsorbents was assessed by thermogravimetric analysis using a TAQ500 instrument. The pre-calcined materials were pretreated in situ at 673 K under flowing N_2 for 1 h. Then CO_2 adsorption was conducted at 573 K and a CO_2 partial pressure of 200 mbar. In multicycle experiments, the samples were regenerated by increasing the temperature to 673 K and purging the cell with pure N_2 for 30 min. The flow was kept constant at 60 mL min^{-1} during the experiment. Further details of the experimental procedure can be found in Chap. 3 (Sect. 3.2.1.1). The unsupported LDH and the hybrids were found to have fast adsorption kinetics at low and intermediate surface coverages reaching 80 % of their equilibrium capacity within 30 min. For practical purposes, the adsorption capacities of the materials were calculated from the weight gain after 120 min of exposure to the CO_2-containing stream although a very slow increase in the adsorption capacity was still observable. Representative CO_2 adsorption profiles for the activated oxidised MWCNTs, the pure LDO, the hybrid 38 wt% MWCNTs (LDO_1) and the hybrid 27 wt% MWCNTs (LDO_2) are displayed in Fig. 4.12.[9] The CO_2 uptake by the nanotubes alone is small compared to the unsupported LDO and its contribution to the adsorption capacity of the hybrids is assumed to be minimal. A maximum in the weight of the oxidised nanotubes is observed over time when exposed to the adsorbate. This profile is likely to be caused by the competition of weak CO_2 adsorption by the MWCNTs and the decomposition of remaining carboxylate groups in the surface or a slow etching of amorphous carbon or defect sites by CO_2. However, based on the cycling stability of the hybrids (see below), the decomposition of the support is expected to be very modest at the temperatures applied [29]. In the presence of water or for intrinsically less stable supports such as carbon nanofibers these background effects may be more significant.

The CO_2 capacities of the activated adsorbents in the first exposure to the adsorptive gas (first-contact capacity) are summarised in Table 4.4. The values presented are an average of at least five measurements of the sample under the same

[8]The CO_2-TPD profile of the pure LDO is further discussed in Chap. 5 (Sect. 5.3.4), where the CO_2 uptake of the sample and the percentage contribution of each desorption peak are given.

[9]Representative adsorption profiles of the pure LDO and LDO_1 at different temperatures and PCO_2 = 200 mbar are presented in Appendix C.3.

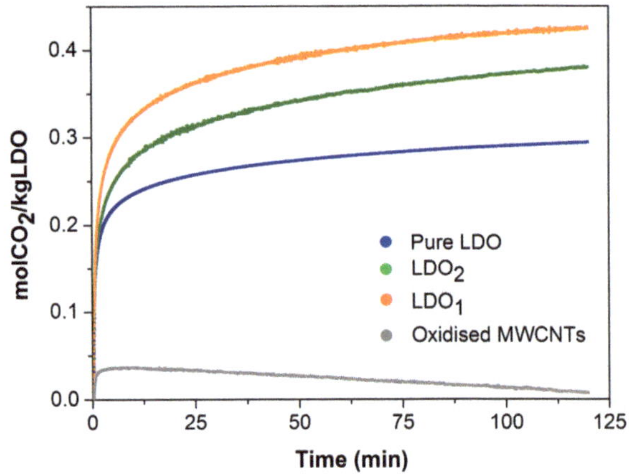

Fig. 4.12 Representative adsorption profiles of pure LDO, LDO_2, LDO_1 and oxidised MWCNTs at 573 K and PCO_2 = 200 mbar. Oxidised MWCNTs curve referred to mol CO_2 kg adsorbent

operating conditions. A good reproducibility was observed for the pure LDH with a standard deviation of 0.02. Conversely, the set of tests conducted with the carbon supported samples gave variable adsorption capacities with standard deviations between 0.06 and 0.15. This is attributed to some heterogeneity in the material on the scale of mass used for TGA (~5 mg). In any case, there is a clear trend showing that the mean adsorption capacity per mass of LDH (intrinsic LDH capacity) under dry conditions increases as the dispersion of LDH and the surface area of the adsorbents increase, i.e. low LDH contents Fig. 4.13. These results are in agreement with previous works [5] that found an enhancement in the intrinsic CO_2 adsorption capacity of LDHs supported on carbon nanofibers although, contrary to the present study, the authors did not find a clear correlation between the capacities and the textural properties of the hybrids.

The adsorption capacities in Table 4.4 and Fig. 4.13 are comparable to results obtained using LDHs supported on zeolites [3] when differences in adsorption temperature and loading are considered but are lower than those reported by Meis et al. when carbon nanofibres are used as support (1.3–2.5 mol CO_2 kg LDH^{-1}) [5]. However, these latter values are not directly comparable to the present work

Table 4.4 First-contact adsorption capacities of activated adsorbents at 573 K and PCO_2 = 200 mbar

Sample name[a]	wt% MWCNTs	mol CO_2 kg LDO[b]	mol CO_2 kg adsorbent[b]
Pure LDO	0	0.30 ± 0.02	0.30 ± 0.02
LDO_4	18	0.36 ± 0.15	0.29 ± 0.15
LDO_2	27	0.38 ± 0.06	0.28 ± 0.06
LDO_1	38	0.42 ± 0.06	0.26 ± 0.06
LDO_1 (larger scale)	39	0.44 ± 0.09	0.27 ± 0.09
$LDO_{0.5}$	67	0.58 ± 0.13	0.19 ± 0.13

[a] LDO_x, where x is the nominal LDO to MWCNTs weight ratio of the sample
[b] Adsorption capacities shown with standard deviations

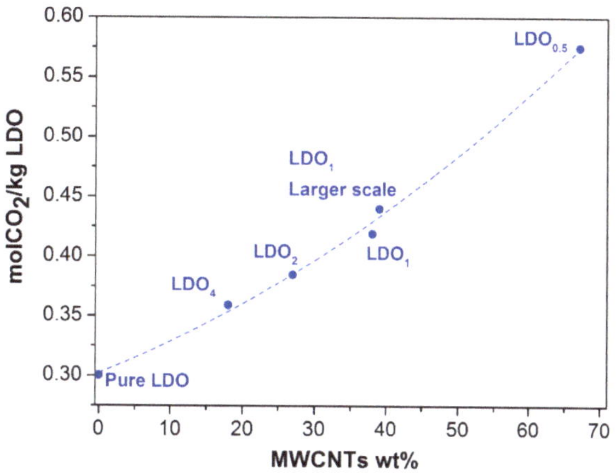

Fig. 4.13 Average first-contact adsorption capacities per mass of LDO at 573 K and PCO_2 = 200 mbar

since the CO_2 adsorption was conducted in the presence of water, which is known to increase significantly the adsorption capacity of LDH [30–32]. Moreover, the adsorption temperature was lower and higher contents of support were used, which increased the specific adsorption capacity.

The regeneration and stability of the adsorbents were assessed by carrying out multiple adsorption-desorption cycles under dry conditions (see Fig. 4.14). The unsupported LDH exhibits a steep fall in the CO_2 adsorption capacity during the first cycles followed by a more gradual decline. The marked initial decay has been reported in previous studies performed under dry conditions and is attributed to a small amount of CO_2 chemisorbed irreversibly over the strongest basic sites of the LDH [31, 33, 34]. For the carbon-supported samples, the profile obtained in multiple cycles is similar to the pure LDH but the severity of the loss of capacity is significantly lower and therefore the stability of LDH is clearly increased by the dispersion on the MWCNTs, Figs. 4.14 and 4.15. It is worth noting that the initial loss in adsorption capacity of the LDHs under dry conditions has been reported to be mitigated by the presence of water, possibly by maintaining the hydroxyl concentration of the surface and/or preventing poisoning through carbonate or coke deposition [31]. For this reason, under hydrated adsorption conditions, the CO_2 capacity of pure and CNF-supported LDHs was reported to be constant over 10 cycles by Meis et al. [5].

The adsorption performance of the carbon hybrids are also compared to that of a commercial adsorbent consisting of 20:80 LDH:α-alumina by weight, Fig. 4.15.[10] The composition and crystalline phases were confirmed by X-ray fluorescence and X-ray powder diffraction respectively (see Appendix C.4). The stability of the alumina-based adsorbent was superior to that of the unsupported LDH in agreement with previous reports [4]. However, the carbon-containing hybrids are markedly

[10]The adsorption capacity at 573 K and PCO_2 = 200 mbar was found to be 0.09 mol CO_2/kg adsorbent (i.e. 0.43 mol CO_2/kg LDO).

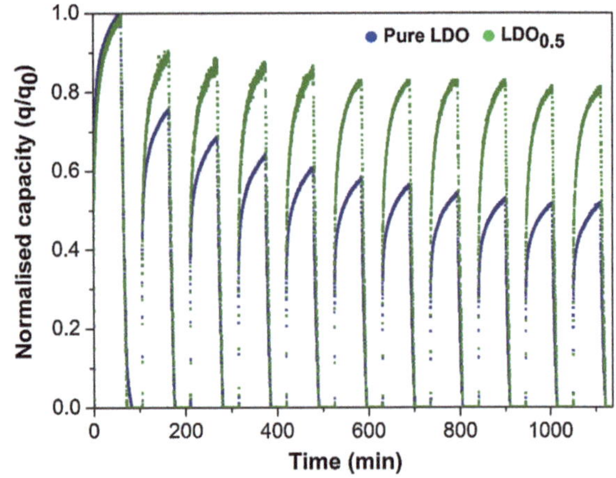

Fig. 4.14 Multicycle profile of the pure LDO and $LDO_{0.5}$ at 573 K

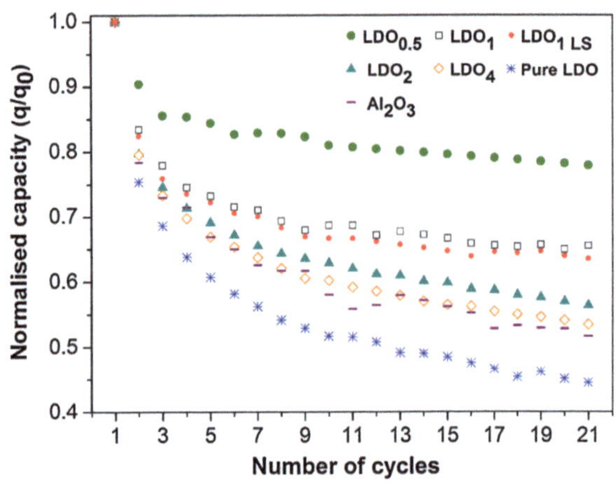

Fig. 4.15 Normalised CO_2 adsorption capacity over 21 adsorption-desorption cycles (adsorption 523 K; desorption 673 K)

more stable even at high total LDH contents, possibly due to the strong network forming ability of MWCNTs in contrast to the relatively low surface area of α-alumina.

The enhancement in the intrinsic adsorption capacity and stability of the LDH supported on MWCNTs may have different origins. It is possible that higher capacities are obtained due to the better accessibility of CO_2 that stems from the improved particle dispersion and increased surface area. Alternatively, the appearance of stacking faults due to the presence of CNTs during the LDH growth may also contribute positively to the adsorption capacity by creating more active sites. In fact, previous studies have suggested that defective sites with low coordination numbers are relevant for the adsorption of CO_2 [5, 6]. Lastly, the dispersion of the LDH on the MWCNT support is likely to mitigate sintering during calcination and regeneration leading to a more stable material.

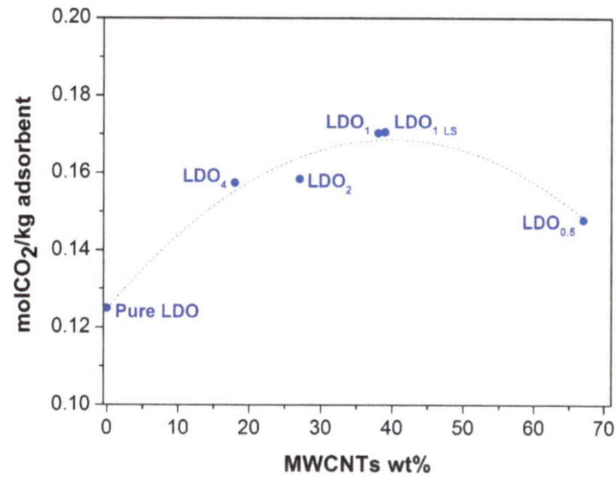

Fig. 4.16 Average adsorption capacities per mass of total adsorbent corresponding to the 21st cycle of the pure LDO and MWCNT hybrids at 573 K and PCO_2 = 200 mbar

So far, the discussion of the beneficial effect of MWCNTs used as supports for LDH has dealt with the increase in the adsorption capacity per mass of LDH (i.e. intrinsic capacity). However, for industrial applications, the adsorption capacity per mass of total adsorbent (i.e. LDH + MWCNTs) has a greater impact. In the first exposure to the adsorbate, the CO_2 capacity per mass of adsorbent was found to increase with the LDH loading as expected from the negligible contribution of the carbon support. As the number of adsorption-desorption cycles increases, the trend in adsorption capacity is reversed since the supported adsorbents are significantly more stable. The adsorption capacity per mass of adsorbent after 21 cycles for the unsupported LDH and the carbon-hybrids are given in Fig. 4.16. It is evident that, when normalising to total mass of adsorbent, there is an optimum LDH loading corresponding to 35–50 wt% MWCNTs.

4.5 Concluding Remarks

In this chapter, layered double hydroxides (LDH) supported on multi-walled carbon nanotubes (MWCNTs) have been studied as adsorbents for carbon dioxide. Precipitation of LDH onto oxidised MWCNTs solubilised in aqueous media is a feasible strategy to synthesise hybrid materials due to a high compatibility of surface charges. It was found that the CO_2 adsorption capacity per mass of LDH increases with the nanotube loading either due to an increment in the structural disorder of the supported materials or to a higher effective surface area, which enhances the adsorbent/adsorbate contact.

Besides the improvement in the intrinsic capacity of the supported LDH derivatives, their stability also increases significantly with MWCNTs addition. Therefore, after adsorption-desorption cycling under dry conditions, an optimum LDH loading (about 60 wt% LDH) was found in terms of the adsorption capacity per total mass of adsorbent. The adsorption performance of the LDH/MWCNT is expected to further improve in the presence of water. Compared to a precompetitive LDH adsorbent supported on alumina, the nanotube hybrids proved to be significantly more stable and require lower loadings of MWCNTs as support, which results in higher capacities per mass of adsorbent.

The adsorption studies were carried out using hybrid samples prepared in batches with yields in the order of milligrams. Additionally, a batch of few grams was prepared to check for any deviations in the properties of the materials introduced by the scale up of the preparation procedure. It was observed that while the morphological properties and the sorption performance of the material produced in larger scale was consistent with the corresponding small batch, the synthesis was time consuming mainly due to longer oxidation and filtration steps. Alternative procedures to oxidise the CNTs should be explored [35]. Likewise, it would be useful to find carbon nanostructures that are compatible with LDHs and that lead to hybrids that are easier to prepare in large scale so that the materials can be tested in breakthrough curve experiments.

References

1. Choi, S., Drese, J. H., & Jones, C. W. (2009). Adsorbent materials for carbon dioxide capture from large anthropogenic point sources. *ChemSusChem, 2*(9), 796–854.
2. Wang, Q., Luo, J., Zhong, Z., & Borgna, A. (2011). CO_2 capture by solid adsorbents and their applications: Current status and new trends. *Energy & Environmental Science, 4*(1), 42–55.
3. Othman, M. R., Rasid, N. M., & Fernando, W. J. N. (2006). Mg–Al hydrotalcite coating on zeolites for improved carbon dioxide adsorption. *Chemical Engineering Science, 61*(5), 1555–1560.
4. Aschenbrenner, O., McGuire, P., Alsamaq, S., Wang, J., Supasitmongkol, S., Al-Duri, B., et al. (2011). Adsorption of carbon dioxide on hydrotalcite-like compounds of different compositions. *Chemical Engineering Research and Design, 89*(9), 1711–1721.
5. Meis, N. N. A. H., Bitter, J. H., & de Jong, K. P. (2009). Support and size effects of activated hydrotalcites for precombustion CO_2 capture. *Industrial and Engineering Chemistry Research, 49*(3), 1229–1235.
6. Meis, N. N. A. H., Bitter, J. H., & de Jong, K. P. (2010). On the Influence and role of alkali metals on supported and unsupported activated hydrotalcites for CO_2 sorption. *Industrial and Engineering Chemistry Research, 49*(17), 8086–8093.
7. Toebes, M. L., van Heeswijk, J. M. P., Bitter, J. H., Jos van Dillen, A., & de Jong, K. P. (2004). The influence of oxidation on the texture and the number of oxygen-containing surface groups of carbon nanofibers. *Carbon, 42*(2), 307–315.
8. Wong, H. S. P., & Akinwande, D. (2011). *Carbon nanotube and graphene device physics*. NY, USA: Cambridge University Press.
9. Krueger, A. (2010). *Carbon materials and nanotechnology*. Cornwall, UK: WILEY-VCH.
10. Othman, M. R., Helwani, Z., Martunus, & Fernando, W. J. N. (2009). Synthetic hydrotalcites from different routes and their application as catalysts and gas adsorbents: A review. *Applied Organometallic Chemistry, 23*(9), 335–346.
11. Georgakilas, V., Gournis, D., Karakassides, M. A., Bakandritsos, A., & Petridis, D. (2004). Organic derivatization of single-walled carbon nanotubes by clays and intercalated derivatives. *Carbon, 42*(4), 865–870.

12. Du, B., & Fang, Z. (2010). The preparation of layered double hydroxide wrapped carbon nanotubes and their application as a flame retardant for polypropylene. *Nanotechnology, 21* (31), 315603.
13. Yang, J.-I., & Kim, J.-N. (2006). Hydrotalcites for adsorption of CO_2 at high temperature. *Korean Journal of Chemical Engineering, 23*(1), 77–80.
14. Climent, M. J., Corma, A., Iborra, S., Epping, K., & Velty, A. (2004). Increasing the basicity and catalytic activity of hydrotalcites by different synthesis procedures. *Journal of Catalysis, 225*(2), 316–326.
15. Fogden, S., Verdejo, R., Cottam, B., & Shaffer, M. S. P. (2008). Purification of single walled carbon nanotubes: The problem with oxidation debris. *Chemical Physics Letters, 460*, 162–167.
16. Ram Reddy, M. K., Xu, Z. P., Lu, G. Q., & Diniz da Costa, J. C. (2006). Layered double hydroxides for CO_2 capture: Structure evolution and regeneration. *Industrial and Engineering Chemistry Research, 45*(22), 7504–7509.
17. Hutson, N. D., Speakman, S. A., & Payzant, E. A. (2004). Structural effects on the high temperature adsorption of CO_2 on a synthetic hydrotalcite. *Chemistry of Materials, 16*(21), 4135–4143.
18. Hu, H., Bhowmik, P., Zhao, B., Hamon, M. A., Itkis, M. E., & Haddon, R. C. (2001). Determination of the acidic sites of purified single-walled carbon nanotubes by acid—base titration. *Chemical Physics Letters, 345*(1–2), 25–28.
19. Verdejo, R., Lamoriniere, S., Cottam, B., Bismarck, A., & Shaffer, M. (2007). Removal of oxidation debris from multi-walled carbon nanotubes. *Chemical Communications, 5*, 513–515.
20. Zhao, H., & Nagy, K. L. (2004). Dodecyl sulfate–hydrotalcite nanocomposites for trapping chlorinated organic pollutants in water. *Journal of Colloid and Interface Science, 274*(2), 613–624.
21. Millange, F., Walton, R. I., & O'Hare, D. (2000). Time-resolved in situ X-ray diffraction study of the liquid-phase reconstruction of Mg–Al–carbonate hydrotalcite-like compounds. *Journal of Materials Chemistry, 10*(7), 1713–1720.
22. Wang, H., Xiang, X., & Li, F. (2010). Facile synthesis and novel electrocatalytic performance of nanostructured Ni-Al layered double hydroxide/carbon nanotube composites. *Journal of Materials Chemistry, 20*(19), 3944–3952.
23. Huang, S., Peng, H., Tjiu, W. W., Yang, Z., Zhu, H., Tang, T., et al. (2010). Assembling exfoliated layered double hydroxide (LDH) nanosheet/carbon nanotube (CNT) hybrids via electrostatic force and fabricating nylon nanocomposites. *The Journal of Physical Chemistry B, 114*(50), 16766–16772.
24. León, M., Díaz, E., Bennici, S., Vega, A., Ordóñez, S., & Auroux, A. (2010). Adsorption of CO_2 on hydrotalcite-derived mixed oxides: Sorption mechanisms and consequences for adsorption irreversibility. *Industrial and Engineering Chemistry Research, 49*(8), 3663–3671.
25. Turco, M., Bagnasco, G., Costantino, U., Marmottini, F., Montanari, T., Ramis, G., et al. (2004). Production of hydrogen from oxidative steam reforming of methanol: I. Preparation and characterization of $Cu/ZnO/Al_2O_3$ catalysts from a hydrotalcite-like LDH precursor. *Journal of Catalysis, 228*(1), 43–55.
26. Rouquerol, F., Rouquerol, J., & Sing, K. (1999). *Adsorption by powders and porous solids*. London, UK: Academic Press.
27. Di Cosimo, J. I., Apesteguía, C. R., Ginés, M. J. L., & Iglesia, E. (2000). Structural requirements and reaction pathways in condensation reactions of alcohols on $MgyAlO_x$ catalysts. *Journal of Catalysis, 190*(2), 261–275.
28. Debecker, D. P., Gaigneaux, E. M., & Busca, G. (2009). Exploring, tuning, and exploiting the basicity of hydrotalcites for applications in heterogeneous catalysis. *Chemistry—A European Journal, 15*(16), 3920–3935.
29. Delpeux, S., Szostak, K., Frackowiak, E., & Béguin, F. (2005). An efficient two-step process for producing opened multi-walled carbon nanotubes of high purity. *Chemical Physics Letters, 404*(4–6), 374–378.

30. Hufton, J. R., Mayorga, S., & Sircar, S. (1999). Sorption-enhanced reaction process for hydrogen production. *AIChE Journal, 45*(2), 248–256.
31. Ding, Y., & Alpay, E. (2000). Equilibria and kinetics of CO_2 adsorption on hydrotalcite adsorbent. *Chemical Engineering Science, 55*(17), 3461–3474.
32. Ram Reddy, M. K., Xu, Z. P., & Diniz da Costa, J. C. (2008). Influence of water on high-temperature CO_2 capture using layered double hydroxide derivatives. *Industrial and Engineering Chemistry Research, 47*(8), 2630–2635.
33. Carvill, B. T., Hufton, J. R., Anand, M., & Sircar, S. (1996). Sorption-enhanced reaction process. *AIChE Journal, 42*(10), 2765–2772.
34. Hutson, N. D., & Attwood, B. C. (2008). High temperature adsorption of CO_2 on various hydrotalcite-like compounds. *Adsorption, 14*(6), 781–789.
35. Menzel, R., Tran, M. Q., Menner, A., Kay, C. W. M., Bismarck, A., & Shaffer, M. S. P. (2010). A versatile, solvent-free methodology for the functionalisation of carbon nanotubes. *Chemical Science, 1*(5), 603–608.

Chapter 5
Layered Double Hydroxides Supported on Graphene Oxide for CO_2 Adsorption

5.1 Introduction

Layered double hydroxide derivatives exhibit adequate CO_2 sorption characteristics in the temperature range from 473 to 723 K which make them very attractive for pre-combustion CO_2 capture applications. However their capacity and stability need to be improved before they are used in large scale. An interesting approach to enhance the performance of layered double hydroxides (LDHs) is to support them on high surface area materials (e.g. zeolites, alumina and carbon nanofibers). Following this strategy adsorbents with high intrinsic capacities (per mass of LDH) have been obtained but the improvement has been observed only with very high loadings of inert support (above ~80 wt%) that would result in large sorption units. In Chap. 4 it was shown that LDH hybrids containing MWCNTs in moderate amounts (below ~50 wt%) show adequate capacities per mass of total adsorbent and are markedly more stable than pure Mg–Al LDHs. In addition, their performance was found to be superior to that of a pre-competitive commercial LDH/Al_2O_3 adsorbent. However, while the results obtained using nanotubes are encouraging, other nanocarbon geometries are expected to be even more suitable. This is the case of graphene oxide (GO) layers which present a 2D geometry that is more obviously compatible with the Mg–Al oxide layers while the surface chemistry is expected to be similarly favourable. In principle a single atomically thin layer of GO may stabilise LDH platelets on each side suggesting a very high mass efficiency, Fig. 5.1. Additionally, the large size of the GO sheets compared to LDH platelets implies that it may be possible to form an open network that allows access to the active adsorption sites.

Preliminary studies on the performance of graphene oxide supported LDHs were carried out in collaboration with the Materials Chemistry group of Prof. Milo Shaffer in the Department of Chemistry at Imperial College London. The adsorbents were synthesised by coprecipitation of $Mg(NO_3)_2$ and $Al(NO_3)_3$ in a colloidal alkaline solution containing the GO nanosheets. Samples in the whole range of GO

D. Iruretagoyena Ferrer, *Supported Layered Double Hydroxides as CO_2 Adsorbents for Sorption-enhanced H_2 Production*, Springer Theses,
DOI 10.1007/978-3-319-41276-4_5

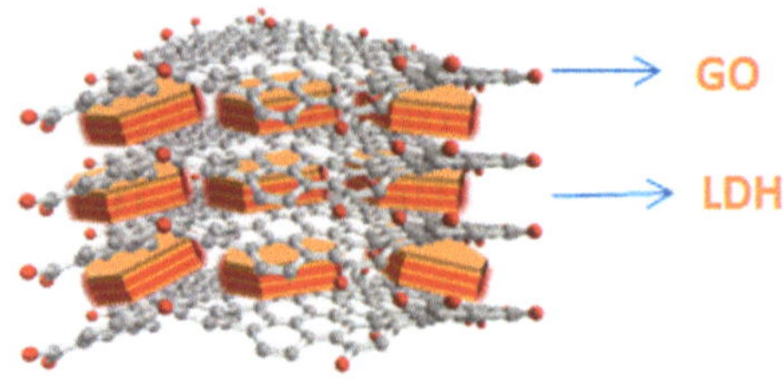

Fig. 5.1 Schematic representation of an ideal LDH/GO adsorbent

contents were prepared, i.e. 0, 7, 12, 33, 83 wt% and pure GO. After aging for 12 h at 333 K, the materials were filtered and dried. For GO contents up to 33 wt%, only the characteristic peaks of LDH were detected and the reflection in the c-direction was found to be broader than that of the pure LDH even for low support loadings. This was attributed to an efficient integration of the GO in the LDH structure which decreased the coherence length in the direction of layer restacking. Above 83 wt% no crystalline LDH was detected while the (002) and (100) reflections of GO were observed, indicating significant restacking of the graphitic material. HRTEM images of the 33 wt% GO sample showed crystalline hexagonal platelets of LDH attached to the GO nanosheets revealing that the strong interaction between the positively charged LDH layers and the negatively charged GO successfully led to the deposition of the crystallites on the support. N_2 physisorption measurements revealed that the surface area of the hybrids increases consistently with the LDH content being higher for the samples with 7 and 12 wt% GO than for the unsupported LDH. This suggests that GO and LDH self-stacking is reduced by the intercalation of LDH and GO layers as indicated by the X-ray diffractograms. The higher surface areas were found to be related to an increase in the number of mesopores as small LDH platelets were incorporated in the hybrid.

The solids were activated prior to CO_2 adsorption measurements at 673 K flowing N_2 to produce supported layered double oxides (LDOs). The activated adsorbents showed higher surface areas than the corresponding as-synthesised samples. It was observed that the initial adsorption capacity per mass of LDO increases by 74 % on adding even a small amount of GO (7 wt%) and increases very modestly at higher support loadings. The enhancement was attributed to the formation of thinner LDH platelets and to their dispersion/stabilisation on a compatible support which leads to an increment in the effective surface area. In addition, continuous adsorption-desorption cycles showed that the hybrid containing 7 wt% GO was significantly more stable than the unsupported LDO. For more details of the preliminary studies on CO_2 adsorption on GO supported LDOs the reader is referred to the corresponding published work which is shown in Ref. [1].

The significant increase in the intrinsic capacity and stability of LDO at relatively low GO loadings is an outstanding result. The present chapter elaborates further on the use of graphene oxide as support for LDOs focusing on the region of low GO loadings (from 1 to 33 wt% GO). The stability of the pure LDO and LDO/GO hybrids was examined over more extended thermal cycling than in the preliminary studies. In addition, the role of GO as support was investigated by temperature-programmed desorption of CO_2 (CO_2-TPD). A better understanding of the adsorption characteristics of the novel materials was also aimed by obtaining CO_2 adsorption isotherms of the materials in the first exposure to the adsorptive gas and after multiple temperature-swing cycles.

The first section of this Sect. 5.2 describes the methodology used to synthesise Mg–Al LDHs and LDH/GO hybrids and the procedure followed to activate the adsorbents before testing. Section 5.3 examines the physicochemical properties of as-synthesised and activated adsorbents by a range of characterisation techniques including TGA, TEM, SEM, XRD, BET and CO_2-TPD. The next Sect. 5.4 focuses on the CO_2 adsorption capacity and multicycle stability of the unsupported and GO supported LDOs. In Sect. 5.5, CO_2 equilibrium data obtained from TGA studies under dry conditions are fitted to Langmuir, Freundlich and Toth isotherms. Finally, a summary of the key findings of the chapter is given in Sect. 5.6.

5.2 Synthesis of the Adsorbents

5.2.1 Unsupported Mg–Al LDHs

Unsupported LDHs, $Mg_{0.6}Al_{0.3}(OH)_2(CO_3)_{0.15}{\cdot}nH_2O$, were prepared via co-precipitation under high supersaturation conditions (see Chap. 2, Sect. 2.4). To prepare 1 g of LDH, an aqueous solution (4.5 mL) of 2 M $Mg(NO_3)_2{\cdot}6H_2O$ and 1 M $Al(NO_3)_3{\cdot}9H_2O$ was added dropwise to an aqueous solution (6.7 mL) of 1. 2 M Na_2CO_3 and 4.7 M NaOH under vigorous stirring. The inorganic salts used in the synthesis were purchased from Sigma-Aldrich. The measured pH of the solution was 10. The resulting white suspension was aged at 333 K for at least 12 h under vigorous stirring using a magnetic bar. Subsequently, the precipitate was filtered using 0.2 μm nylon membranes and thoroughly washed with DI water (0.5 L per 1 g of solid) at 333 K. Samples were dried for 24 h at 393 K in a recirculating oven. Typically 1 to 5 g of unsupported LDHs were prepared per batch.

5.2.2 Nanostructured LDH/GO Hybrids

Graphene oxide powder (NanoInnova Technologies) was dispersed in a solution containing 1.2 M Na_2CO_3 and 4.7 M NaOH from which the LDH was co-precipitated. The amount of GO added reflected the desired loading of GO in the hybrid. The resulting suspension was aged, filtered, washed and dried as described above for the unsupported LDH. Five different LDH/GO hybrids were synthesised with nominal weight ratios 99/1 (LDH_{99}), 20/1 (LDH_{20}), 7/1 (LDH_7), 3/1 (LDH_3) and 2/1 (LDH_2), using, respectively, 10, 50, 143, 333 and 500 mg of graphene oxide at constant LDH precursor concentration.

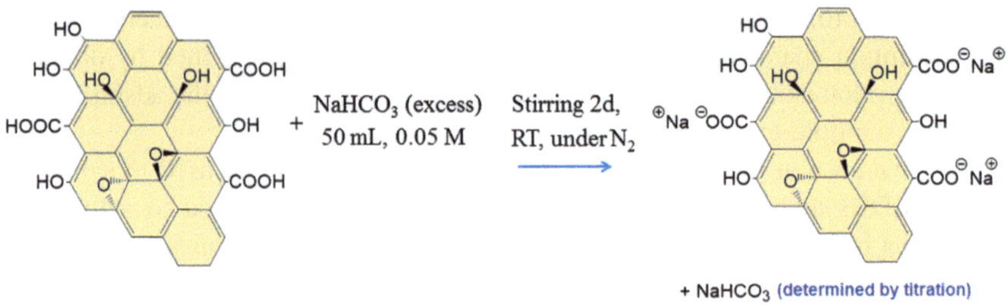

Fig. 5.2 Determination of carboxylic acid groups on the graphene oxide by reverse titration

5.2.3 *Activation of Pure LDH and LDH/GO Hybrids*

Prior to CO_2 adsorption measurements, the materials were activated by calcination in situ flowing 20 mL min^{-1} of N_2 at 673 K for 4 h to produce the corresponding unsupported LDO and the hybrids LDO_{99}, LDO_{20}, LDO_7, LDO_3 and LDO_2. The activation was performed in situ in the TGA to avoid irreversible loss in capacity due to CO_2 captured from exposure to the atmosphere. For characterisation purposes, the as-synthesised samples were activated ex situ flowing 100 mL min^{-1} of N_2 at 673 K for 4 h using a quartz cylinder (ID = 4 cm, L = 60 cm) placed in a horizontal furnace (Lenton). It has been widely reported that a calcination temperature of 673 K produces LDH derivatives with an adequate balance between surface area and basic sites, which maximises their CO_2 capacities and favours reversible adsorption (further details can be found in Chap. 2, Sect. 2.4.1) [2, 3].

5.2.4 *Titration of Graphene Oxide*

The carboxylic acid group concentration present in the graphene oxide was estimated by adapting the modified Boehm titration procedure established for the MWCNTs (Chap. 4, Sect. 4.2.4) [4, 5]. 50 mg of GO were stirred in 50 mL of 0. 05 M $NaHCO_3$ aqueous solution under nitrogen for 2 days at room temperature. After filtration of the mixture and washing with deionized water, the combined filtrate and washings were added to 50 mL of 0.05 M HCl. The excess of HCl in the solution was titrated with 0.05 M NaOH to neutral, as monitored by a pH meter. Figure 5.2 shows a schematic representation of the procedure followed.

The concentration of carboxylic acid groups on the graphene oxide was estimated to be 1.23 mmol g^{-1}, which corresponds to a surface charge density of 22 (μmol e^-) m^{-2}.[1] For the LDH, the extra positive charge due to the substitution of

[1]The surface charge density was calculated using the surface area of the dried graphene oxide sample measured by N_2 physisorption (55 m^2 g^{-1}). The surface area of the GO sample is relatively low compared to that of the completely exfoliated GO, probably due to restacking during drying.

Al^{3+} is equivalent to 0.33 electrons per mol and is contained within the area of the octahedral unit, i.e. 8.067 Å^2 [6]. Therefore, the excess surface charge density of LDH is approximately 6.7 (μmol e^+) m^{-2}, which is on the same order of magnitude that the surface charge density of the GO. Hence, the idea that LDH tends to nucleate on and remains associated with the graphene oxide surface is reasonable.

5.3 Characterisation

The as-synthesised and activated adsorbents were analysed by the following physicochemical techniques: inductively coupled plasma-optical emission spectroscopy (ICP-OES), X-ray diffraction (XRD), transmission electron microscopy (TEM), scanning electron microscopy (SEM), thermogravimetric analysis (TGA), N_2 physisorption at 77 K and temperature-programmed desorption of CO_2 (CO_2-TPD). The detailed description of the equipment and the test procedures was given in Chap. 3 (Sect. 3.4).

5.3.1 Composition of the Adsorbents

The compositions of the pure LDH and LDH/GO hybrids are given in Table 5.1. The average Mg/Al ratio of the as-synthesised adsorbents, determined by ICP, was found to be 2.2 ± 0.06 which is close to the intended ratio of 2. Sodium was detected only at trace levels since the materials were thoroughly washed to remove NaOH and Na_2CO_3 used during the synthesis. However, it can be observed that the sodium content while small, increases with the GO loading. This is caused by the attraction between the sodium ions and the highly negatively GO sheets during the LDH coprecipitation. In fact, it was found that a hydrothermally treated GO (i.e. stirring the GO in water for 12 h at 333 K and pH 10) contained 4 wt% of sodium after it was thoroughly washed during the filtration step (2 L of water per 150 mg of

Table 5.1 Composition of the as-synthesised adsorbents

Sample name[a]	wt% GO Nom.	wt% GO actual	Mg/Al (mo mol^{-1})	wt% Na
Pure LDH	0	0	2.2	0.02
LDH_{99}	1	2	2.3	0.02
LDH_{20}	5	4	2.2	0.07
LDH_7	13	9	2.2	0.05
LDH_3	25	17	2.2	0.40
LDH_2	33	20	2.1	0.54

[a]LDH_x, where x is the nominal LDH to GO weight ratio of the sample

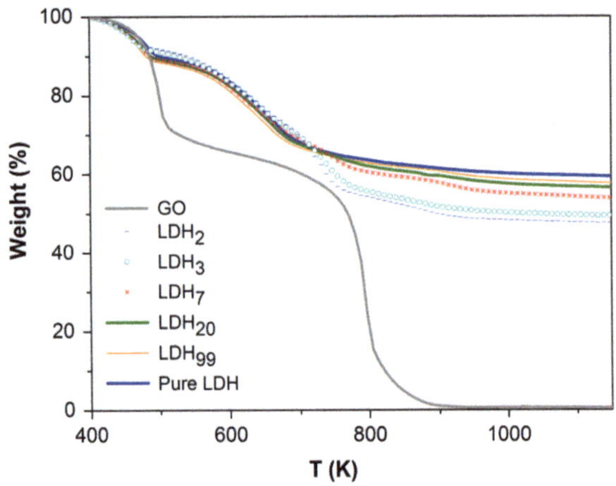

Fig. 5.3 Thermograms of pure LDH, GO and LDH/GO hybrid samples

GO). This shows that it is not feasible to reduce the residual sodium in the samples below the levels in Table 5.1. It is assumed that the adsorption properties of the pure LDH and the LDH hybrids are not influenced by this low level of alkali content as it is likely to be mainly associated with the GO.

The TGA patterns of the adsorbents are shown in Fig. 5.3. The unsupported LDH exhibited three stages of weight loss corresponding to its dehydration (below ~483 K), partial dehydroxylation and decarbonation (between ~483 and ~733 K) and further decarbonation (above ~733 K). The TGA of the pure GO presents a weight loss at around 500 K due to the decomposition of functional groups and the oxidation debris, and a weight loss at around 750 K which has been attributed to the combustion of the (damaged) graphitic regions [7]. In the hybrids, the weight loss at high temperature correlates with the GO content. Therefore, the actual GO weight percentage in the LDH/GO materials can be estimated from the residue of each sample (see Chap. 3, Sect. 3.4.6). The TGA residue of the pure LDH is composed of ~60 wt% mixed solid oxides, while the residue of GO can be considered negligible. The nominal and actual weight loadings are very close for the lower GO content samples (LDH_{99} and LDH_{20}). However, the higher GO loading adsorbents have considerably less GO than expected (Table 5.1). This is in agreement with previous observations showing that the loss of oxidative debris during filtering becomes more significant as the GO content in the sample increases [1]. In the present study a brown liquid was observed passing through the 0.2 μm filter membrane during the washing step for the LDH_2 and LDH_3 hybrids, which is ascribed to this debris. The oxidative debris associated with the graphene oxide consists mainly of fragments of polyaromatic hydrocarbons functionalised with acid oxides, which are generated in the original oxidation process. This debris acts as a surfactant to stabilise aqueous dispersions of GO. However under basic conditions these acidic groups become conjugate bases increasing their water solubility and allowing their filtration [7].

5.3.2 Crystal Morphology

The crystallinity of the adsorbents was investigated by XRD (Fig. 5.4). All the LDH containing samples display the characteristic reflections of 2D hydrotalcite-like materials with carbonates in the interlayer (JCPDS No. 14-191) and can be indexed accordingly [8]. The pure GO shows reflection lines at 11.3° and 42.3° assigned to the (002) and (100) planes. The sharp GO peak at 11.3°, corresponding to an interlayer distance of 0.78 nm, completely disappears after hydrothermal treatment and washing. Instead, a broad peak appears at 27.1° corresponding to an interlayer distance of 0.33 nm. This broad peak is associated to disordered restacking of exfoliated, partially oxidised graphene sheets remaining after the elimination of the more oxidised debris, by filtration under basic conditions [7]. However, this reflection is only observed at very high loadings of GO (e.g. LDH/GO hybrid with 91 wt% GO, see Fig. 1e in [1]) and therefore is absent from the patterns of the GO-supported LDHs presented in this chapter. For the hybrids, the intensity of the LDH reflections and crystallite size decreased slightly with the amount of GO (Fig. 5.4, Table 5.2). In general, the dispersion (surface area) was found to correlate with the crystallite size (see Tables 5.2 and 5.3).

After calcination, all the samples (LDH and hybrids) show broad reflection peaks ascribed to the diffraction planes of periclase (MgO, JCPDS No. 45-946); the aluminium compounds are thought to be well dispersed or in an amorphous phase, Fig. 5.5. The calcination of the pure GO led to diffraction lines corresponding to graphitic (002) and (100) lattice planes whereas in the XRD pattern of the hydrothermally treated GO only a peak representative of graphite (002) is clearly distinguished. In the hybrids this reflection was observed as a weak signal only at relatively high loadings of support (i.e. LDO_3 and LDO_2) indicating that LDH does not prevent all restacking of the graphene sheets for these concentrations. The crystallite sizes of the activated adsorbents are given in Table 5.2.

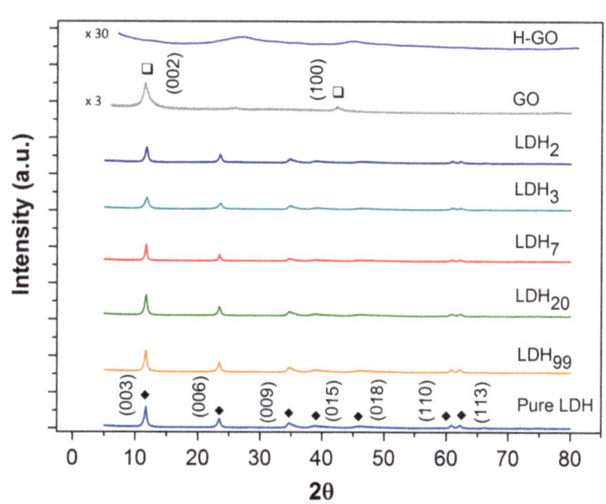

Fig. 5.4 Representative XRD diffraction patterns of pure LDH, LDH/GO hybrids, GO and hydrothermally treated GO (H-GO). Characteristic reflections of LDH (♦) and GO (❑)

Table 5.2 Crystallite size of the as-synthesised and activated adsorbents

Sample name[a]	Crystallite size (003) (nm)	Sample name[a]	Crystallite size (200) (nm)
Pure LDH	21	Pure LDO	3.2
LDH_{99}	18	LDO_{99}	3.2
LDH_{20}	19	LDO_{20}	3.1
LDH_7	16	LDO_7	3.1
LDH_3	11	LDO_3	2.9
LDH_2	16	LDO_2	2.8

[a]LDH_x and LDO_x where x is the nominal LDH to GO weight ratio of the sample

Table 5.3 Textural properties of as-synthesised and activated adsorbents

Sample name[a]	As-synthesised adsorbents			Sample name[a]	Activated adsorbents		
	SBET ($m^2\ g^{-1}$)	Vpore ($cm^3\ g^{-1}$)	Dpore (nm)		SBET ($m^2\ g^{-1}$)	Vpore ($cm^3\ g^{-1}$)	Dpore (nm)
Pure LDH	112	0.46	12	Pure LDO	223	0.51	10
LDH_{99}	116	0.62	16	LDO_4	205	0.83	16
LDH_{20}	106	0.54	15	LDO_2	199	0.62	13
LDH_7	113	0.54	16	LDO_1	219	0.64	11
LDH_3	143	0.47	10	LDO_3	170	0.54	10
LDH_2	114	0.55	17	$LDO_{0.5}$	157	0.66	15
GO	55	0.06	6	GO	81	0.15	6

[a]LDH_x and LDO_x, where x is the nominal LDH to GO weight ratio of the sample

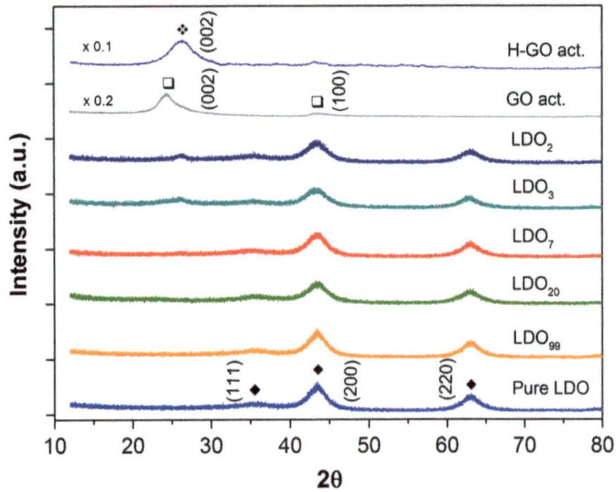

Fig. 5.5 Representative XRD diffraction patterns of activated adsorbents. Characteristic reflections of LDO (♦) and calcined GO (graphene ❑ and graphite ❖)

TEM images of the unsupported and supported LDHs show small, crystalline, hexagonal platelets of LDH, Fig. 5.6. As expected the shape and average size of the LDH particles (~20 nm) in all the as-synthesised adsorbents were found to be

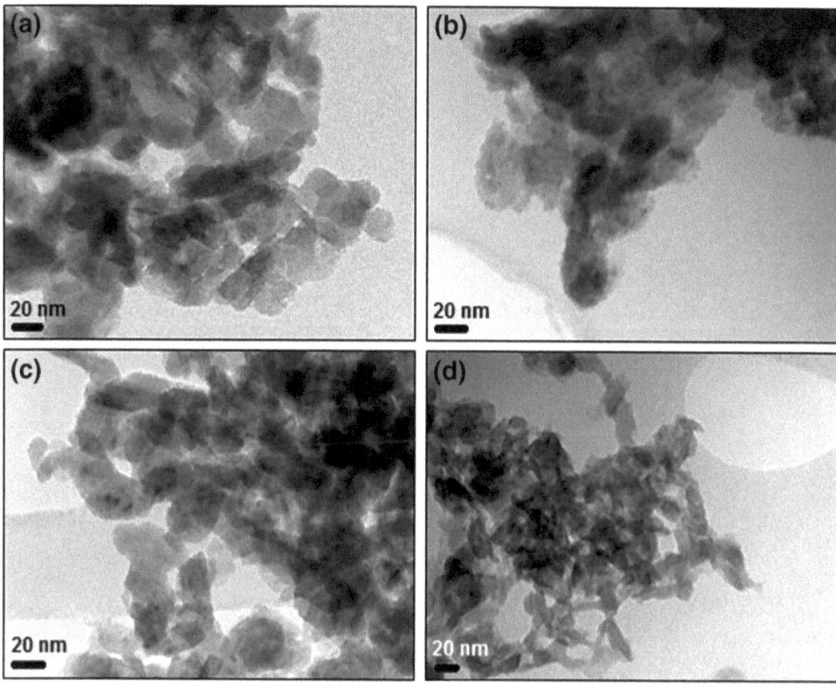

Fig. 5.6 Representative TEM images of pure LDH (**a**), LDH_{20} (**b**), LDH_7 (**c**) and LDH_3 (**d**)

similar. No GO nanosheets were observed in the TEM images of the hybrids. It is assumed that for these low-medium loadings, the GO nanosheets are well integrated with the LDH particles. TEM images reported previously for hybrids with higher GO contents show the presence of LDH particles supported on GO nanosheets with flake like morphologies, [1].

In agreement with the TEM analysis, SEM images revealed that the surface topologies of the LDH containing samples were similar, Fig. 5.7. In addition, no morphological differences were apparent at the microscale between the as-synthesised and the activated adsorbents. This observation seems to be consistent with other research which found that Zn-Al LDHs maintained the morphology of the original precursor after thermal treatment at 723 and 773 K [9, 10].

5.3.3 Textural Properties

The textural properties of the adsorbents were determined by N_2 physisorption. All the as-synthesised and activated materials exhibited type IV isotherms with H3 hysteresis loops [11] as observed for the LDH/MWCNT hybrids (see Chap. 4,

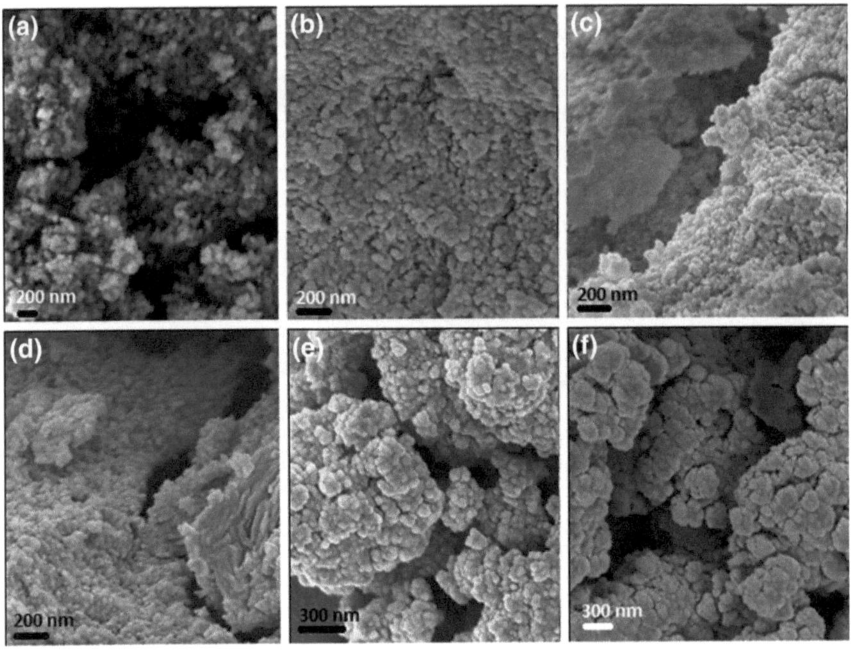

Fig. 5.7 Representative SEM images of pure LDH (**a**), LDH_{20} (**b**), LDH_7 (**c**) LDH_3 (**d**, **e**) and LDO_3 (**f**)

Sect. 4.3.3). Representative isotherms are shown in Fig. 5.8.[2] The BET surface areas and the pore diameters and volumes obtained by the BJH method are given in Table 5.3. The surface area of the unsupported LDH was found to be high compared to other reported values for similar methods [12].[3] This is assumed to be caused by improved stirring and control of the rate of addition of the solution containing the Mg and Al nitrates. The fact that the pure LDH and the hybrids show similar porosity suggests that higher loadings of GO are required to notice marked changes in the dispersion of the as-synthesised hybrids as those observed in preliminary studies [1]. Upon calcination, the surface area of the adsorbents increases but the increment is slightly lower for the hybrids with high content of GO (LDO_3 and LDO_2). This is likely to be caused by GO restacking which becomes more significant as the amount of intercalating hydroxide layers decreases.

[2]Representative N_2 physisorption isotherms of the LDH/GO hybrids that are not presented in this chapter can be found in Appendix D.1.

[3]See the surface area of the pure LDH reported in Chap. 4 and Ref. [1].

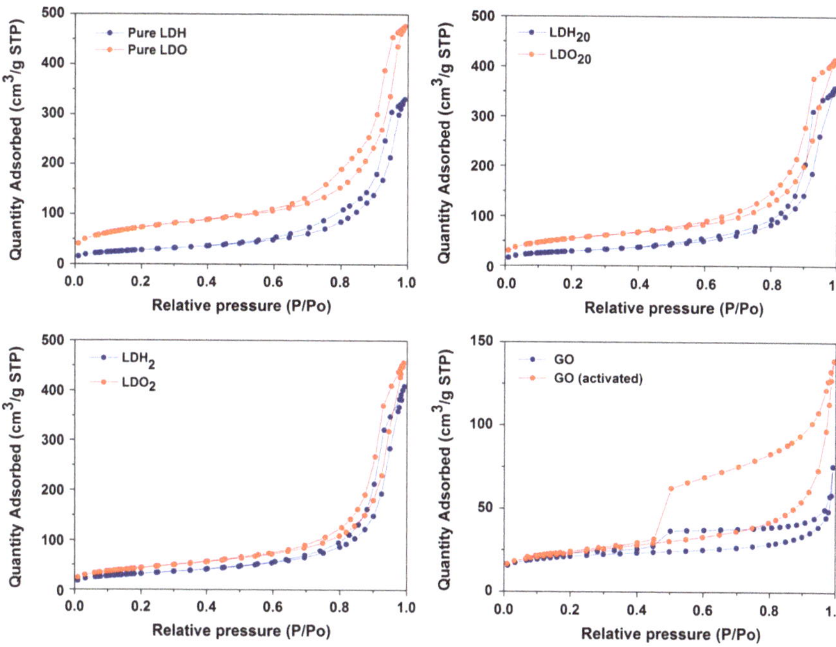

Fig. 5.8 Representative N_2 physisorption isotherms of as-synthesised and activated adsorbents

5.3.4 Temperature-Programmed Desorption of CO_2

Temperature-programmed desorption (TPD) of CO_2 was used to investigate the influence of GO on the density and strength of the basic sites of the activated adsorbents. The desorption profile of the unsupported LDO and the hybrids containing 4 wt% of GO (LDO_{20}) and 20 wt% of GO (LDO_2) are presented in Fig. 5.9. Similarly to the CO_2-TPD of the pure LDO in Fig. 4.11 (Chap. 4), the present adsorbents show three overlapping peaks with maxima at 423 K (α), 543 K (β) and 813 K. The high temperature state can be deconvoluted into two states centred at 813 K (γ) and 963 K (δ). These basic sites of increasing strength are assigned to the formation of bicarbonates, bidentate and monodentate carbonates, respectively [13, 14] and correspond to desorption energies in the range 100–200 kJ mol^{-1} (assuming a pre-exponential factor of 10^{13} and negligible readsorption) [15, 16]. Weaker adsorption states and physisorbed CO_2 are not observed in the TPD due to the initial purging. After the pretreatment at 673 K, the low and intermediate energy states are the main contributors to the CO_2 uptake of the LDO (adsorbents) determined by TPD. The distribution of basic sites in the hybrid containing a relatively low amount of GO, LDO_{20}, is very close to that of the pure LDO. The presence of 20 wt% of GO in the LDO did not modify the distribution of the low and intermediate energy states but the desorption peak at high temperature (γ) is

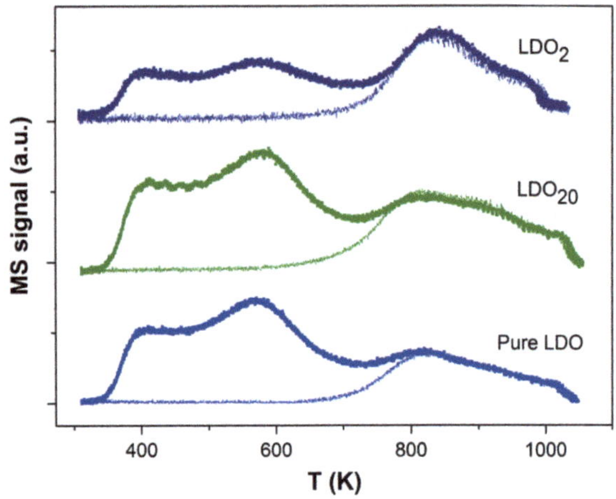

Fig. 5.9 CO_2-TPDs of pure LDO, LDO_{20} and LDO_2 following pretreatment at 673 K and adsorption at 313 K (*dark lines*). The blanks were subjected to pretreatment only (*light lines*)

slightly larger (Fig. 5.9 and Table 5.4). This peak corresponds to the evolution of remaining carbonates from the LDH structure and to the thermal decomposition of GO, Fig. 5.10.

The CO_2 uptake of the unsupported LDO, LDO_{20} and LDO_2 is given in Table 5.5. These values were obtained by subtracting the corresponding blank, i.e. the CO_2 evolved during the temperature program from the activated samples after pretreatment that were not exposed to CO_2 adsorption (Fig. 5.9 and Table 5.5). The similar density and distribution of basic sites of the unsupported and supported LDOs revealed by CO_2-TPD suggests that the presence of GO, in the range of loadings studied, does not modify significantly the nature of the adsorption sites available after calcination at 673 K. This does not preclude the possibility that higher loadings of GO and/or alternative preparation methods may result in significantly modified sites.

As mentioned in Chap. 4, the use of intermediate adsorption temperatures and low regeneration gradients are generally more favourable for temperature-swing applications. Consequently, in this study, adsorption and desorption temperatures of 573 and 673 K were used to assess the multicycle stability of the adsorbents and is reported in Sect. 5.4. Under CO_2 adsorption at low pressure, mainly the desorption state centred at 543 K and part of the state at 813 K are cyclically populated and

Table 5.4 Percentage contribution of each desorption peak: CO_2-TPD

Sample[a]	α (423 K)	β (543 K)	γ (813 K)	δ (963 K)
Pure LDO	13	55	28	4
LDO_{20}	12	49	33	6
LDO_2	9	40	49	2

The deconvolution of the TPD profiles was carried out using Fityk 0.8.9
[a]LDO_x, where x is the nominal LDO to GO weight ratio of the sample
[b]Bicarbonates (α), bidentates (β), monodentates (γ and δ)

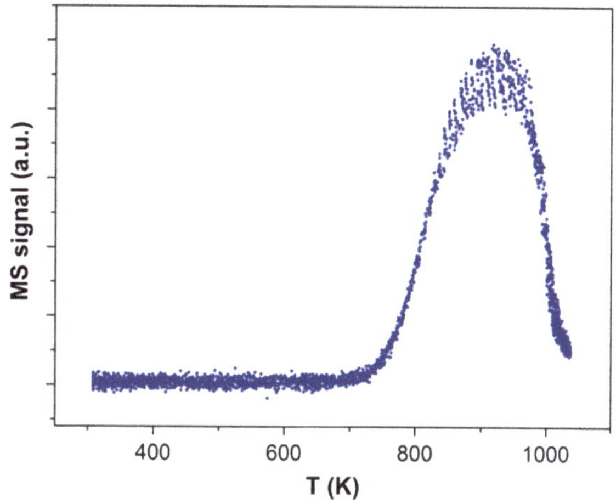

Fig. 5.10 CO_2-TPD of hydrothermally treated GO. Similar desorption profiles were obtained for the original GO and the corresponding blanks (i.e. subjected to pretreatment only)

Table 5.5 Density of basic sites: CO_2-TPD

Sample name[a]	wt% GO actual	Total CO_2 evolved (μmol g^{-1})	CO_2 evolved[b] (minus blank) (μmol g^{-1})	CO_2 evolved[b] (minus blank) (μmol g LDO^{-1})	CO_2 evolved[b] (minus blank) (μmol m^{-2} ads)
Pure LDO	0	600	440	440	1.96
LDO_{20}	4	690	410	430	2.07
LDO_2	20	710	360	450	2.29

[a]LDO_x, where x is the nominal LDO to GO weight ratio of the sample
[b]Values obtained by subtracting the corresponding blank from the total CO_2 evolved

depopulated, while the low temperature sites are nearly empty. However, as the CO_2 pressure during adsorption at 573 K is increased, weaker states become populated and ultimately physisorption contributes to the adsorption capacity [2, 17].

5.4 CO_2 Adsorption Capacity and Multicycle Stability

The CO_2 adsorption performance of the adsorbents was assessed using a thermogravimetric analyser (Perkin Elmer, TGA4000). An amount ($\sim$15 mg) of sample was calcined in situ at 673 K under flowing N_2 for 4 h. Then CO_2 adsorption was conducted at 573 K and a CO_2 partial pressure of 200 mbar. In multicycle tests, the adsorption step was carried out at 573 K for 1 h flowing the premixed CO_2 gas and the desorption step was performed at 673 K during 30 min flowing nitrogen. The flow rate was kept at 20 ml min^{-1} during the experiment. Further details of the experimental procedure can be found in Chap. 3 (Sect. 3.2.1.1).

All the activated adsorbents showed fast kinetics at 573 K requiring 30 min to achieve ca. 90 % of their equilibrium capacity. For practical purposes the values

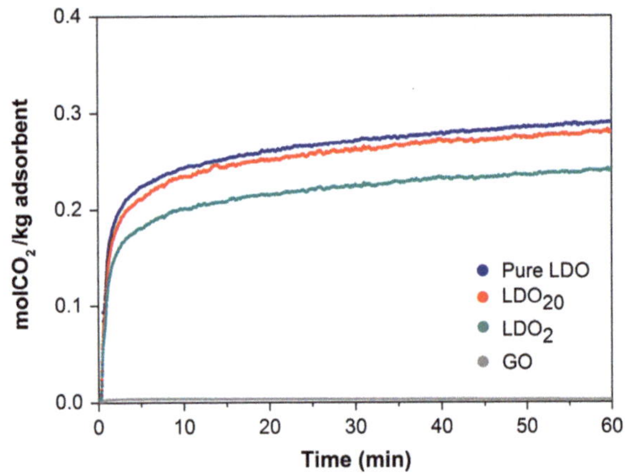

Fig. 5.11 Representative adsorption profiles of pure LDO, LDO_{20}, LDO_2 and GO, at 573 K and PCO_2 = 200 mbar

here reported correspond to those obtained after 60 min of CO_2 exposure at 200 mbar (Fig. 5.11). The adsorption capacities in the first exposure to the adsorptive gas (first-contact capacity) are given in Table 5.6. Under the operating conditions used, the contribution of GO to the CO_2 capacity of the hybrids was negligible. The intrinsic adsorption capacities (i.e. per mass of LDO) of the pure LDO and the LDO/GO hybrids are the same within experimental error. This is consistent with the TPD results which indicate that the density of available sites after activation is essentially the same for the unsupported and supported LDOs. It should be noted that the CO_2 capacity determined at 200 mbar is less than the total uptake in TPD since only a fraction of the CO_2 is in equilibrium as shown by isotopic exchange measurements [14]. As expected, the adsorption capacity normalised with respect to the total mass of adsorbent decreases slightly with the small amount of GO in the hybrid.

Continuous adsorption-desorption cycles were carried out to assess the regeneration and stability of the adsorbents under dry conditions. The CO_2 capacity of the pure LDO and the LDO/GO hybrids drops markedly during the first cycles and becomes stable gradually, Fig. 5.12. The fact that the initial loss of capacity decreases as the GO content in the sample increases suggests that rapid thermal

Table 5.6 First-contact adsorption capacities of the adsorbents at 573 K and PCO_2 = 200 mbar

Sample name[a]	wt% GO actual	mol CO_2 kg LDO^{-1b}	mol CO_2 kg ads^{-1b}
Pure LDO	0	0.29 ± 0.01	0.29 ± 0.01
LDO_{99}	2	0.28 ± 0.03	0.28 ± 0.03
LDO_{20}	4	0.29 ± 0.03	0.28 ± 0.03
LDO_7	9	0.32 ± 0.03	0.29 ± 0.03
LDO_3	17	0.30 ± 0.02	0.25 ± 0.02
LDO_2	20	0.30 ± 0.03	0.24 ± 0.03

[a]LDO_x, where x is the nominal LDO to GO weight ratio of the sample
[b]The adsorption capacities are shown with standard errors

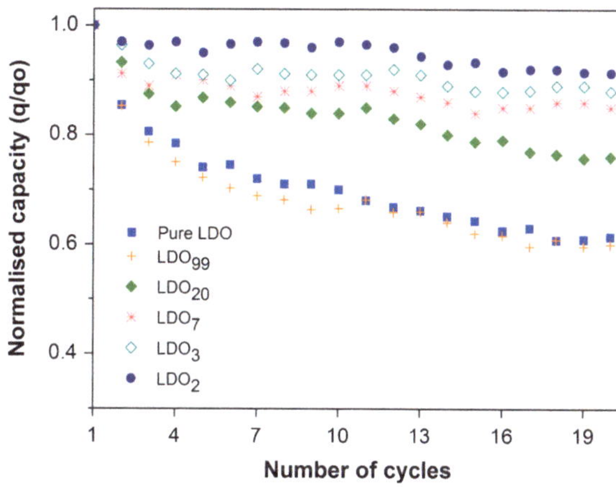

Fig. 5.12 Normalised CO_2 adsorption capacity over 20 adsorption-desorption cycles (adsorption 523 K; desorption 673 K)

sintering takes place followed by a gradual thermal deactivation. Some CO_2 irreversibly chemisorbed on the fresh material may also contribute to the observed profiles as suggested by other studies dealing with dry CO_2 adsorption on LDO containing adsorbents at high temperatures (573–753 K) [17–19]. The presence of water during the adsorption and/or desorption steps is expected to mitigate the initial decay in capacity observed [19, 20].

The stability of the pure LDO is markedly enhanced by the addition of graphene oxide and increases with the GO content in the hybrid, Fig. 5.12. For example, the adsorption capacity of the unsupported LDO after 20 cycles is 60 % of its initial value whereas the hybrid with the highest content of GO studied (LDO_2) reduced its initial capacity by only 10 %. All the hybrids exhibited a minor loss of weight during the first cycles and the weight remained constant thereafter. The loss of mass is related to the partial thermal decomposition of the GO and therefore it becomes more significant for high GO loadings. Remarkably, this does not affect the multicycle stability of the adsorption capacity over the time scale of the study as observed in Fig. 5.12.

To investigate the factors that cause the improvement of the stability of the hybrids, some adsorbents were characterised before and after multicycle testing by XRD and CO_2-TPD. Materials recovered from different runs under the same experimental conditions were used for this purpose. The crystallite sizes of the samples were found to increase after multiple adsorption-desorption cycling, Table 5.7. Although the difference in crystallite size between the fresh and spent materials is relatively small, it seems that graphene oxide helps to prevent the growth of the LDO particles.

The first-contact and multicycle CO_2-TPDs of the pure LDO and the LDO_3 hybrid are presented in Fig. 5.13. The profile of the pure LDO after cycling shows a significant decrease in the number of low and intermediate energy states while the strongest sites are only slightly affected. This leads to a decrease in the heterogeneity of the adsorption sites on the unsupported LDO, Fig. 5.13a. On the

Table 5.7 Crystallite size of fresh and spent activated adsorbents

Sample name[a]	Crystallite size (200) (nm) fresh	Crystallite size (200) (nm) spent
Pure LDO	3.2	4.1
LDO_7	3.1	4.3
LDO_3	2.9	3.4
LDO_2	2.8	3.1

[a]LDO_x, where x is the nominal LDO to GO weight ratio of the sample

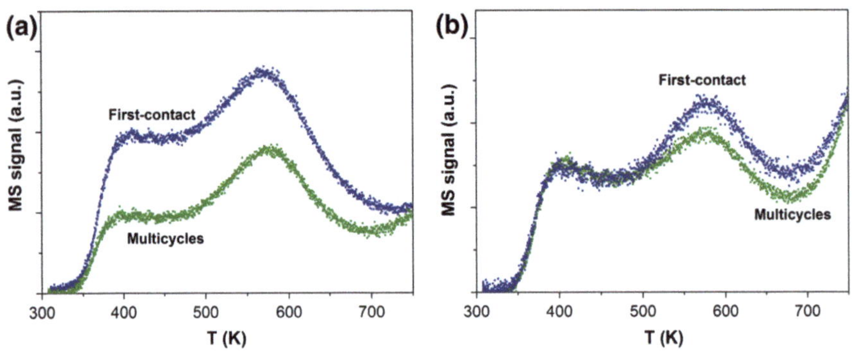

Fig. 5.13 CO_2-TPDs of **a** pure LDO (first-contact and multicycles), **b** LDO_3 (first-contact and multicycles)

contrary, in the presence of graphene oxide there is a small loss of intermediate and high energy states in similar proportions and therefore the initial distribution of adsorption sites is better preserved, Fig. 5.13b.

As mentioned in Sect. 5.2.2, all the LDO/GO hybrids were prepared using graphene oxide powder supplied by NanoInnova Technologies. In order to assess the influence of different sources of GO on the capacity and the stability of the hybrids, LDO_{20} (i.e. with 4 wt% GO actual) was synthesised using GO purchased from ACS Material in flake and water dispersed forms. The adsorption capacities of the resulting hybrids are given in Table 5.8 and their multicycle profiles are presented in Fig. 5.14. It is apparent that there are no substantial differences in the adsorption performance of the materials which confirms the robustness of the preparation method used. However, the ACS samples consistently led to slightly higher intrinsic capacities than the NanoInnova adsorbent while the latter appeared to be slightly more stable.

The overall results of this chapter unambiguously show that the first-contact capacity per mass of LDO does not change significantly in hybrids with low loadings of graphene oxide. However, the thermal stability of the unsupported LDO is markedly enhanced by the addition of these modest amounts of GO. While the improvement in stability is in agreement with the results from the preliminary tests, the intrinsic capacities of the LDO/GO hybrids here presented contrast with the

Table 5.8 First-contact adsorption capacities of LDO_{20} hybrids at 573 K and PCO_2 = 200 mbar

Sample name[a]	wt% GO Actual	wt% Na	mol CO_2 kg LDO^{-1}	mol CO_2 kg ads^{-1}
$LDO_{20,}$ GO powder (NanoInnova Tech.)	4	0.07	0.29 ± 0.03	0.28 ± 0.03
$LDO_{20,}$ GO flakes (ACS material)	4	0.04	0.32 ± 0.03	0.30 ± 0.03
LDO_{20}, GO water dispersion (ACS material)	4	0.01	0.33 ± 0.03	0.32 ± 0.03

[a]LDO_x, where x is the nominal LDO to GO weight ratio of the sample

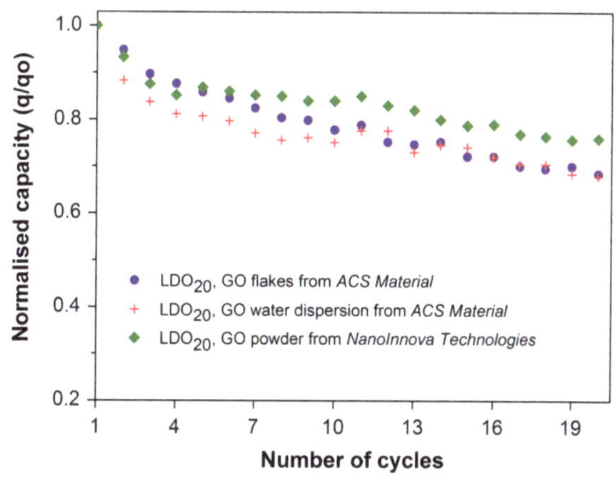

Fig. 5.14 Normalised CO_2 adsorption capacity of LDO_{20} hybrids over 20 adsorption-desorption cycles (adsorption 523 K; desorption 673 K)

relatively high values obtained for similar GO loadings [1]. It is likely that the enhanced capacities achieved in the preliminary studies were associated with the presence of sodium remaining from the synthesis of the hybrid adsorbents. Although it was found that sodium was absent based on single-point EDS measurements, possibly the distribution of sodium was inhomogeneous and therefore difficult to detect. ICP as used here is intrinsically more sensitive and gives an average concentration of sodium in the samples.[4] The discrepancy between both studies motivated to investigate in detail the influence of residual sodium on the CO_2 adsorption performance of the unsupported and GO supported LDOs, and is addressed in Chap. 6.

Table 5.9 compares the adsorption capacity of LDO/GO hybrids with values reported for LDO/MWCNTs (Chap. 3), LDO/Al_2O_3 (Chap. 3) and LDO/CNFs [12]. It is apparent that a significant increase in the intrinsic adsorption capacity of

[4]The composition of the samples of the preliminary studies was measured by ICP (after the work was published). The values are presented in Appendix D.3. The content of sodium in the samples was found to be relatively high (1.2–1.7 wt% Na).

Table 5.9 Comparison with other LDO adsorbents

Authors[b]	Adsorbent	Support (wt%)	mol CO_2 kg LDO^{-1} First-contact	mol CO_2 kg ads^{-1} First-contact	mol CO_2 L ads^{-1a} First-contact	Relative capacity at 20 cycles (%)	mol CO_2 L ads at 20 cycles
Meis et al. [12]	LDO	0	0.10	0.10	0.10	Unknown	Unknown
Meis et al. [12]	LDO/CNF	90	1.30	0.13	0.11	Unknown	Unknown
Chapter 3	LDO	0	0.28	0.28	0.29	45	0.13
Chapter 3	LDO/Al_2O_3	80	0.43	0.09	0.19	52	0.10
Chapter 3	LDO/MWCNTs (LDO_1)	38	0.42	0.26	0.19	65	0.12
Present chapter	LDO	0	0.29	0.29	0.30	60	0.18
Present chapter	LDO/GO (LDO_{20})	4	0.29	0.28	0.29	75	0.22
Present chapter	LDO/GO (LDO_2)	20	0.30	0.24	0.26	91	0.24

[a]The particle density of the adsorbents was determined by helium picnometry (Sect. 3.4.9) except for the LDO/CNF hybrid, which was estimated considering the density of the pure CNF [21]

[b]Experimental conditions

Meis et al. [12]. PCO_2 = 50 mbar, 523 K, wet conditions, TSA regeneration at 773 K under dry conditions

Chapter 3 PCO_2 = 200 mbar, 573 K, dry conditions, TSA regeneration at 673 K under dry conditions pure LDO can be obtained by using relatively high contents of support (see values for alumina and CNF hybrids). However, this leads to low adsorption capacities per total volume of adsorbent which may limit the commercial use of these materials. In Chap. 3, the adsorption performance of LDO/MWCNTs hybrids covering a wide range of MWCNTs loadings (33–82 wt% MWCNTs) was investigated using equivalent operating conditions to those of the present study. It was found that after adsorption-desorption cycling there is an optimum LDO loading (about 40 wt% MWCNTs) in terms of adsorption capacity per total mass of adsorbent. Graphene oxide appears to be especially effective at supporting LDO compared to MWCNTs since a better compromise between the adsorption capacity per total volume of adsorbent and the amount of support required is achieved for the same number of cycles. This high mass efficiency can be attributed to the compatibility of GO and LDO in terms of charge and geometry.

5.5 CO_2 Adsorption Isotherms

The accurate description of the adsorption equilibrium is crucial for the prediction of the performance of large scale units. However, due to differences in the experimental conditions, in the composition (i.e. Mg/Al ratio and presence of promoters) and/or in the method used to prepare the LDH derivatives, there are some inconsistencies in the models reported in the literature to fit CO_2 adsorption data. Langmuir isotherms have been extensively used in their pure form [18, 19, 22] although in some cases modifications have been made to account for chemical complexation [23] or for the presence of two different adsorption sites [24]. On the other hand, the Freundlich model has been applied mainly at low and intermediate CO_2 partial pressures [22, 25, 26]. In this section, the CO_2 adsorption isotherms of the pure LDO and the hybrids LDO_{20} (4 wt% GO actual) and LDO_3 (17 wt% GO actual) are presented. The experiments were conducted in a TGA under dry conditions as described below. The equilibrium data was fitted to Langmuir and Freundlich isotherms in their pure forms. In addition, the use of more advanced models was assessed by fitting the experimental data to the Toth isotherm.

5.5.1 Thermogravimetric Measurements

Thermogravimetric analysis was used to obtain adsorption isotherms under dry conditions by means of two TGA instruments (TAQ500 and TGA4000). The samples were activated (calcined) in situ under flowing N_2 at 673 K for 4 h. The adsorption isotherms were determined using premixed gases: 2, 10, 15, 20, 50, 80, 99 % (v/v) CO_2 in Ar. For first-contact isotherms a fresh sample was activated and then subsequently exposed to a specific CO_2 mixture at 573 K until the adsorption profile was essentially steady. In multicycle experiments a mixture of 20 % CO_2/N_2

was used during the adsorption cycle and the desorption step was performed increasing the temperature to 673 K under flowing nitrogen for 30 min. After near-constant capacity was obtained, the adsorption isotherm was measured continuing the adsorption-desorption cycles but increasing successively the concentration of CO_2 in the adsorption step using the premixed gases. Further details of the experimental procedure can be found in Chap. 3 (Sect. 3.2.1.1).

5.5.2 Fitting of the Adsorption Isotherms

The adsorption capacity of the samples was taken as the average of at least three measurements under the same operating conditions. The individual values were used to calculate the corresponding standard errors of the mean. For the adsorption isotherms after multiple cycles, the standard error of the mean was calculated from the adsorption capacity after twenty cycles obtained in at least three different experiments. This error was assumed to be the same for the adsorption capacities obtained at different CO_2 partial pressures. The experimental adsorption isotherms were fitted to Langmuir (Eq. 5.1), Freundlich (Eq. 5.2) and Toth (Eq. 5.3) models applying a nonlinear least square method using OriginPro 8.6.

$$q = mbp_{CO_2}/(1 + bp_{CO_2}) \tag{5.1}$$

$$q = kp_{CO_2}^{1/n} \tag{5.2}$$

$$q = mbp_{CO_2}/\left(1 + (bp_{CO_2})^t\right)^{1/t} \tag{5.3}$$

Direct fitting of adsorption data using nonlinear regression is usually preferred as it avoids the problems of changing error distribution and biased parameters associated with linear transformations [27].

5.5.3 Simple Adsorption Isotherms

The adsorption isotherms obtained by TGA at 573 K for the pure LDO, LDO_{20} and LDO_3 were fitted to the Langmuir and Freundlich models. The experimental data and the fittings for the first-contact adsorption (equilibrium capacity in the first exposure to the adsorptive gas after activation) are shown in Fig. 5.15. The associated derived parameters are given in Table 5.10. First-contact adsorption by pure LDO, LDO_{20} and LDO_3 is better described by the Freundlich rather than the Langmuir isotherm, although the difference is only just statistically significant for the range of conditions studied. The values of the Freundlich gas-solid interaction parameter, k, are similar for both materials indicating that the presence of GO does

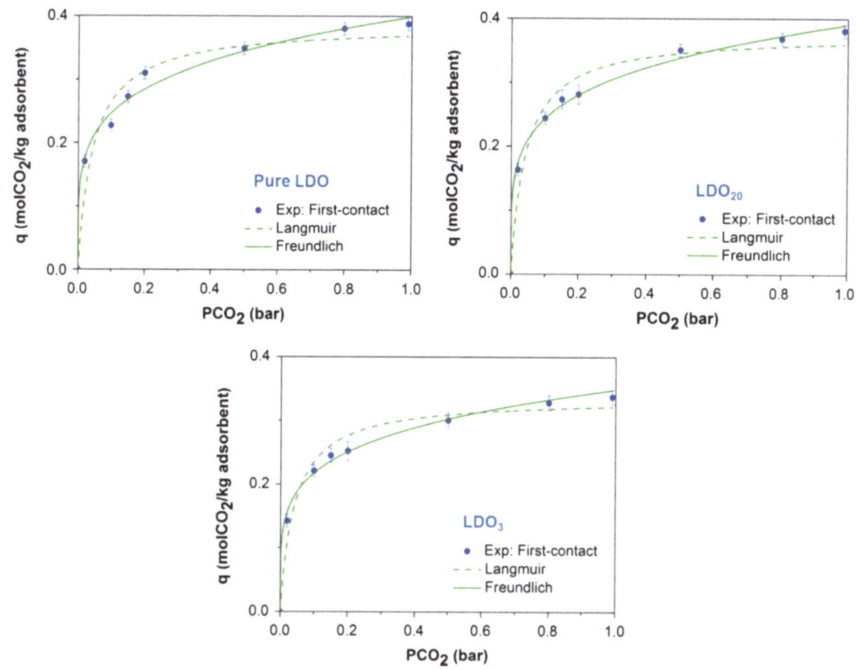

Fig. 5.15 First-contact adsorption isotherms at 573 K under dry conditions from TGA measurements for pure LDO, LDO_{20} (4 wt% GO actual) and LDO_3 (17 wt% GO actual); shown with standard errors

Table 5.10 Parameters for first-contact adsorption isotherms at 573 K from TGA under dry conditions

Adsorbent	Langmuir isotherm[a]			Freundlich isotherm[a]		
	m (mol kg^{-1})	b (bar^{-1})	R^2	k (mol kg^{-1})	n (–)	R^2
Pure LDO	0.39 ± 0.02	21.7 ± 6.5	0.95	0.40 ± 0.01	4.78 ± 0.38	0.99
LDO_{20}	0.37 ± 0.02	23.5 ± 6.3	0.96	0.39 ± 0.01	4.86 ± 0.25	0.99
LDO_3	0.34 ± 0.01	23.5 ± 5.7	0.97	0.35 ± 0.01	4.89 ± 0.33	0.99

[a]The parameters are shown with the standard errors obtained from the fittings

not modify significantly the nature of the adsorption sites. These results are consistent with the CO_2-TPD findings reported above in this chapter (Sect. 3.4), which show that the fresh LDO and LDO/GO hybrids exhibit similar density and distribution of basic sites.

As shown in Sect. 5.4, the multicycle stability of the adsorbents increases with the content of graphene oxide in the sample (i.e. $LDO_3 > LDO_{20} > LDO$). The adsorption isotherms of these materials after being exposed to 20 sorption cycles show that only LDO_3 is described by the Freundlich model whereas the LDO and

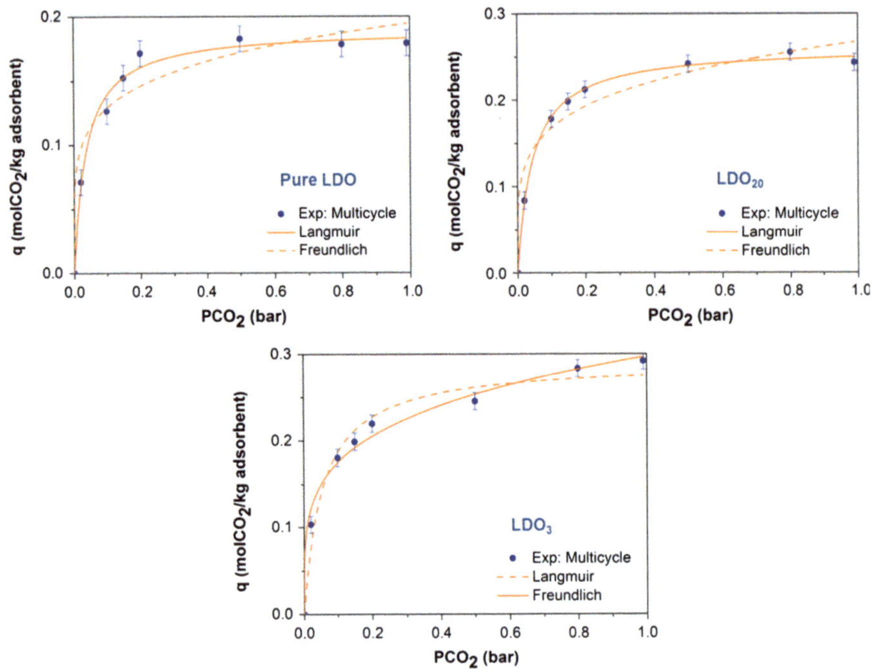

Fig. 5.16 Multicycle adsorption isotherms at 573 K under dry conditions from TGA measurements for pure LDO, LDO_{20} (4 wt% GO actual) and LDO_3 (17 wt% GO actual); shown with standard errors

LDO_{20} isotherms were found to fit better to the Langmuir model, Fig. 5.16 and Table 5.11. This indicates that upon adsorption-desorption temperature cycling the heterogeneity of the adsorption sites tends to disappear rendering a more ideal uniform surface. However, the presence of sufficient amounts of GO as in LDO_3 helps to maintain the initial distribution of the basic strength for a longer number of temperature-swing cycles, which is in agreement with the CO_2-TPD results presented in Sect. 5.4 (Fig. 5.13).

Table 5.11 Parameters for multicycle adsorption isotherms at 573 K from TGA under dry conditions

Adsorbent	Langmuir isotherm[a]			Freundlich isotherm[a]		
	m (mol kg^{-1})	b (bar^{-1})	R^2	k (mol kg^{-1})	n (–)	R^2
Pure LDO	0.19 ± 0.01	27.2 ± 4.3	0.99	0.20 ± 0.01	5.58 ± 1.28	0.93
LDO_{20}	0.26 ± 0.00	22.2 ± 1.5	0.99	0.27 ± 0.01	4.91 ± 0.91	0.95
LDO_3	0.29 ± 0.01	17.9 ± 3.9	0.97	0.30 ± 0.01	4.34 ± 0.34	0.99

[a]The parameters are shown with the standard errors obtained from the fittings

5.5.4 Toth Isotherm

The gradual change of LDO based adsorbents during cycling from a wide distribution of adsorption sites towards a homogeneous surface suggests that it may be convenient to use more sophisticated models to describe the CO_2 adsorption equilibrium. An obvious choice is the Toth model which includes an additional fitting parameter, t (Eq. 5.3). As t approaches 1 the Toth isotherm reduces to the Langmuir model while for values deviating from unity the expression describes adequately the heterogeneity of the material.

The fittings of the first-contact and multicycle experimental data to the Toth model for the pure LDO, LDO_{20} and LDO_3 are presented in Fig. 5.17, Tables 5.12 and 5.13. It can be observed that the Toth equation successfully describes all the experimental data. The fitted t values for the first-contact adsorption isotherms are smaller than 1 but tend to unity after successive adsorption-desorption cycles. As expected, the change is more marked for LDO and LDO_{20} compared to LDO_3 which has a higher concentration of GO. The value of t for the LDO_3 sample is far from unity after cycling confirming that the heterogeneity of adsorption sites is better maintained as the amount of GO in the material increases. Consequently, the value of t is a function of the GO content and of the number of adsorption-desorption cycles to which the samples have been subjected.

5.5.5 Estimation of the Heat of Adsorption

To calculate the heat of adsorption of CO_2 over pure LDO and LDO_{20}, the isotherms in the first contact to the adsorptive and after multiple cycles were obtained at 623 K. The experiments were carried out using a thermogravimetric analyser (Perkin Elmer, TGA 4000), regenerating in situ the samples at 673 K under flowing N_2 for 30 min. The stability of the samples during 20 adsorption-desorption cycles is presented in Fig. 5.18. The loss of adsorption capacity was found to be slightly more marked than the observed when the adsorption takes place at 573 K. However, the stability of the adsorbents increases with the addition of GO in agreement with the trends presented in Sect. 5.4 (Fig. 5.12).

The first-contact and spent adsorption isotherms of the pure LDO and LDO_{20} are presented in Fig. 5.19, Tables 5.14 and 5.15. The CO_2 uptakes at the different partial pressures are lower than the corresponding values obtained at 573 K as expected. Similar to the isotherms obtained at 573 K, the equilibrium data of the first exposure to CO_2 at 623 K for both materials is better described by the Freundlich model and tends to Langmuir after consecutive cycles. The corresponding fittings to the Toth isotherm are presented in Appendix D.2.

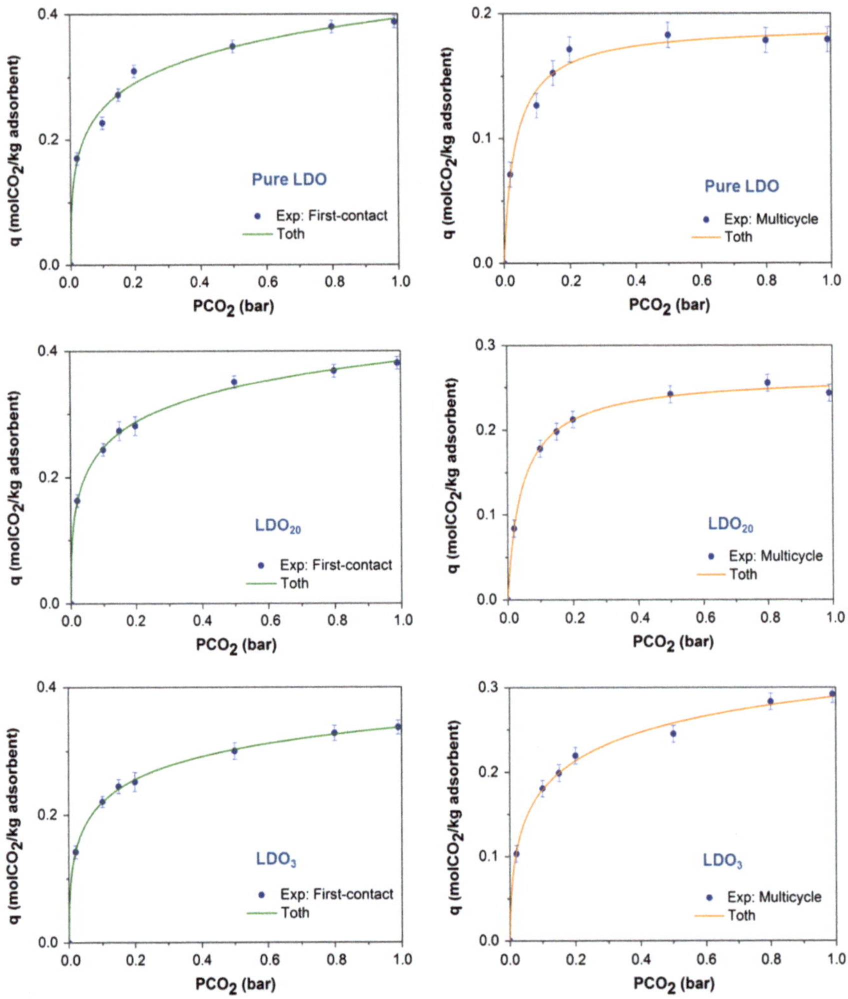

Fig. 5.17 First-contact and multicycle Toth isotherms at 573 K under dry conditions from TGA measurements for pure LDO, LDO_{20} (4 wt% GO actual) and LDO_3 (17 wt% GO actual); shown with standard errors

Table 5.12 Toth parameters for first-contact isotherms at 573 K from TGA under dry conditions

Adsorbent	Toth isotherm[a]			
	m (mol kg^{-1})	b (bar^{-1})	*t* (–)	R^2
Pure LDO	1.14 ± 1.37	10,322 ± 71,537	0.17 ± 0.16	0.99
LDO_{20}	0.88 ± 0.29	3394 ± 6652	0.21 ± 0.06	0.99
LDO_3	0.63 ± 0.09	1260 ± 1077	0.25 ± 0.04	0.99

[a]The parameters are shown with the standard errors obtained from the fittings

Table 5.13 Toth parameters for multicycle isotherms at 573 K from TGA under dry conditions

Adsorbent	Toth isotherm[a]			
	m (mol kg^{-1})	b (bar^{-1})	t (–)	R^2
Pure LDO	0.19 ± 0.01	28.6 ± 10.9	0.96 ± 0.25	0.99
LDO_{20}	0.27 ± 0.01	24.8 ± 4.1	0.92 ± 0.10	0.99
LDO_3	0.47 ± 0.11	185.5 ± 209.9	0.33 ± 0.10	0.99

[a]The parameters are shown with the standard errors obtained from the fittings

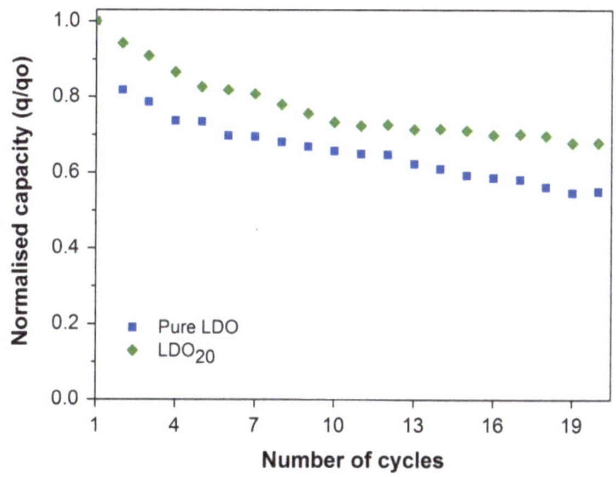

Fig. 5.18 Normalised CO_2 adsorption capacity of pure LDO and LDO_{20} over 20 adsorption-desorption cycles (adsorption 623 K; desorption 673 K)

The heats of adsorption, Q_{ads}, were obtained from the isotherms at 573 and 623 K at a constant uptake of 0.01 mol kg^{-1}.[5] The Langmuir model was used in the analysis even for the first-contact adsorption since it predicts the Henry's law region in which the isosteric heat was calculated. The values of heat of adsorption of the pure LDO and LDO_{20} (first-contact and multicycle) are given in Table 5.16. It is apparent that all the values are of the same order of magnitude which further supports the idea that GO does not modify significantly the nature of the adsorption sites. The fact that the isosteric heats are relatively low suggests that physisorption contributes to the total CO_2 uptake of the materials. Similar Henry's law isosteric heats have been reported in the literature for alkali promoted LDOs (8–29 kJ mol^{-1}) [18, 19, 28].

Differences between the first-contact and multicycle isosteric heats of adsorption were expected due to the observed changes in the heterogeneity of the adsorbents upon cycling. However, the isosteric heats here presented are broad estimates and only provide a correct idea of the order of magnitude of the heats involved. The relatively small variations in the heats of adsorption are within the experimental error (±17 kJ mol^{-1}) and therefore cannot be interpreted further. A detailed study using different temperature gradients and a larger number of low CO_2 partial

[5]The equation used to calculate the isosteric heat of adsorption can be found in Sect. 2.2.1.1 of Chap. 2.

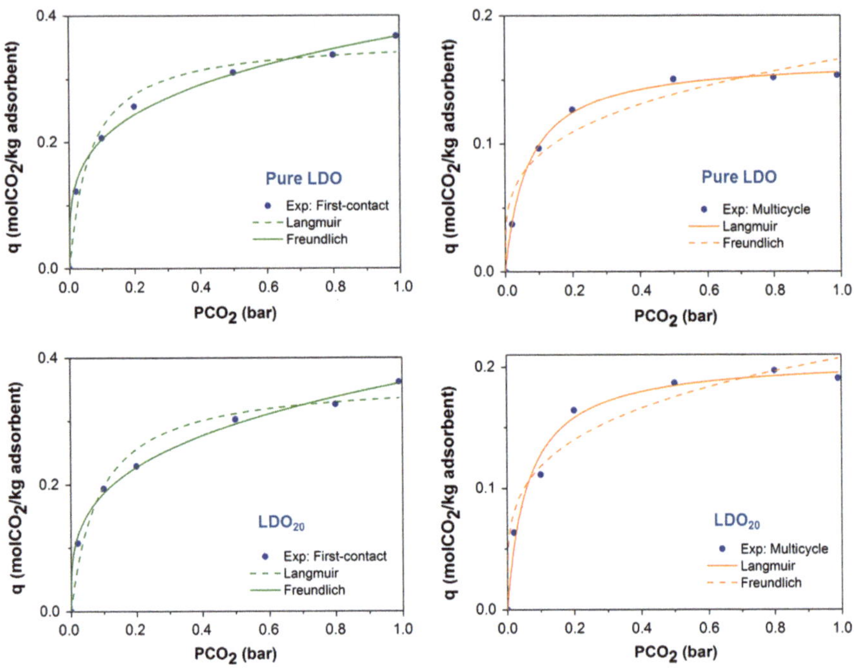

Fig. 5.19 First-contact and multicycle isotherms at 623 K under dry conditions from TGA measurements for pure LDO and LDO_{20} (4 wt% GO actual); shown with standard errors

Table 5.14 Parameters for first-contact adsorption isotherms at 623 K from TGA under dry conditions

Adsorbent	Langmuir isotherm[a]			Freundlich isotherm[a]		
	m (mol kg^{-1})	b (bar^{-1})	R^2	k (mol kg^{-1})	n (–)	R^2
Pure LDO	0.36 ± 0.02	15.5 ± 4.02	0.97	0.37 ± 0.01	3.90 ± 0.22	0.99
LDO_{20}	0.36 ± 0.02	11.7 ± 3.19	0.97	0.36 ± 0.01	3.49 ± 0.17	0.99

[a]The parameters are shown with the standard errors obtained from the fittings

Table 5.15 Parameters for multicycle adsorption isotherms at 623 K from TGA under dry conditions

Adsorbent	Langmuir isotherm[a]			Freundlich isotherm[a]		
	m (mol kg^{-1})	b (bar^{-1})	R^2	k (mol kg^{-1})	n (–)	R^2
Pure LDO	0.17 ± 0.01	14.5 ± 0.91	0.99	0.17 ± 0.01	3.81 ± 0.75	0.95
LDO_{20}	0.21 ± 0.01	15.9 ± 3.20	0.98	0.21 ± 0.01	4.08 ± 0.71	0.96

[a]The parameters are shown with the standard errors obtained from the fittings

Table 5.16 Isosteric heats of adsorption

Adsorbent	First-contact	Multicycle
	Q_{ads} (kJ mol^{-1})	Q_{ads} (kJ mol^{-1})
Pure LDO	48	25
LDO_{20}	43	34

pressures is needed to obtain accurate values. Generally, the use of large temperature gradients results in more precise values but may cause changes in the structure of the adsorbents which need to be taken into account.

5.6 Concluding Remarks

In this chapter the preparation of hybrids containing LDH and GO was studied as a strategy to improve the adsorption performance of layered double hydroxide derivatives. The coprecipitation of $Mg(NO_3)_2$ and $Al(NO_3)_3$ in an alkaline solution containing the dispersed GO was proven to lead to the deposition of the positively charged LDH crystallites on the negative GO nanosheets. As a result, LDH and GO layers are intercalated in the hybrids.

The thermal stability of activated LDHs (i.e. LDOs) is markedly enhanced by the addition of modest amounts of graphene oxide (below ~20 wt%) causing the CO_2 adsorption capacity of the LDO/GO hybrids to be better preserved upon cycling. On the other hand, the CO_2 adsorption capacity per mass of LDO varies only slightly in the presence of these modest amounts of GO. CO_2-TPD suggests that the presence of graphene oxide does not modify significantly the nature and the density of the CO_2 adsorption sites. Compared to other materials such as MWCNTs or alumina, GO appears to be especially effective as a support since a better compromise between adsorption capacity per total volume of adsorbent and the amount of support used is achieved.

The first-contact adsorption data for the pure LDO and the GO hybrids is better described by the Freundlich isotherm but after multiple temperature-swing cycles the isotherms tend to fit to the Langmuir model as the surface becomes more uniform. The use of GO as a support significantly enhances the thermal stability of the LDOs helping to maintain the number and heterogeneity of the adsorption sites for a longer number of temperature- swing cycles. CO_2-TPDs of the pure LDO after activation exhibit a broad distribution of adsorption sites which decreases after consecutive cycles due to a marked loss of high energy states. On the contrary, in the presence of graphene oxide there is a small loss of intermediate and strong basic sites in similar proportions and therefore the initial distribution is better preserved. The gradual homogenisation of the surface of the adsorbents is correctly predicted by the Toth model in which the t parameter is a function of the number of cycles for fixed solid composition and operating conditions. The magnitudes of the isosteric heats of adsorption estimated for the pure LDO and the GO containing hybrids suggest that physisorption contributes to the adsorption capacities of the materials.

References

1. Garcia-Gallastegui, A., Iruretagoyena, D., Gouvea, V., Mokhtar, M., Asiri, A. M., Basahel, S. N., et al. (2012). Graphene oxide as support for layered double hydroxides: Enhancing the CO_2 adsorption capacity. *Chemistry of Materials, 24*(23), 4531–4539.
2. Hutson, N. D., Speakman, S. A., & Payzant, E. A. (2004). Structural effects on the high temperature adsorption of CO_2 on a synthetic hydrotalcite. *Chemistry of Materials, 16*(21), 4135–4143.
3. Ram Reddy, M. K., Xu, Z. P., Lu, G. Q., & Diniz da Costa, J. C. (2006). Layered double hydroxides for CO_2 capture: Structure evolution and regeneration. *Industrial and Engineering Chemistry Research, 45*(22), 7504–7509.
4. Hu, H., Bhowmik, P., Zhao, B., Hamon, M. A., Itkis, M. E., & Haddon, R. C. (2001). Determination of the acidic sites of purified single-walled carbon nanotubes by acid–base titration. *Chemical Physics Letters, 345*(1–2), 25–28.
5. Verdejo, R., Lamoriniere, S., Cottam, B., Bismarck, A., & Shaffer, M. (2007). Removal of oxidation debris from multi-walled carbon nanotubes. *Chemical Communications, 5*, 513–515.
6. Zhao, H., & Nagy, K. L. (2004). Dodecyl sulfate–hydrotalcite nanocomposites for trapping chlorinated organic pollutants in water. *Journal of Colloid and Interface Science, 274*(2), 613–624.
7. Rourke, J. P., Pandey, P. A., Moore, J. J., Bates, M., Kinloch, I. A., Young, R. J., et al. (2011). The real graphene oxide revealed: stripping the oxidative debris from the graphene-like sheets. *Angewandte Chemie, 50*(14), 3173–3177.
8. Millange, F., Walton, R. I., & O'Hare, D. (2000). Time-resolved in situ X-ray diffraction study of the liquid-phase reconstruction of Mg–Al–carbonate hydrotalcite-like compounds. *Journal of Materials Chemistry, 10*(7), 1713–1720.
9. Latterini, L., Elisei, F., Aloisi, G. G., Costantino, U., & Nocchetti, M. (2002). Space-resolved fluorescence properties of phenolphthalein-hydrotalcite nanocomposites. *Physical Chemistry Chemical Physics, 4*(12), 2792–2798.
10. Turco, M., Bagnasco, G., Costantino, U., Marmottini, F., Montanari, T., Ramis, G., et al. (2004). Production of hydrogen from oxidative steam reforming of methanol: I. Preparation and characterization of $Cu/ZnO/Al_2O_3$ catalysts from a hydrotalcite-like LDH precursor. *Journal of Catalysis, 228*(1), 43–55.
11. Rouquerol, F., Rouquerol, J., & Sing, K. (1999). *Adsorption by powders & porous solids*. London, UK: Academic Press.
12. Meis, N. N. A. H., Bitter, J. H., & de Jong, K. P. (2009). Support and size effects of activated hydrotalcites for precombustion CO_2 capture. *Industrial and Engineering Chemistry Research, 49*(3), 1229–1235.
13. Debecker, D. P., Gaigneaux, E. M., & Busca, G. (2009). Exploring, tuning, and exploiting the basicity of hydrotalcites for applications in heterogeneous catalysis. *Chemistry—A European Journal, 15*(16), 3920–3935.
14. Di Cosimo, J. I., Apesteguía, C. R., Ginés, M. J. L., & Iglesia, E. (2000). Structural requirements and reaction pathways in condensation reactions of alcohols on MgyAlOx catalysts. *Journal of Catalysis, 190*(2), 261–275.
15. Redhead, P. A. (1962). Thermal desorption of gases. *Vacuum, 12*(4), 203–211.
16. Falconer, J. L., & Schwarz, J. A. (1983). Temperature-programmed desorption and reaction: Applications to supported catalysts. *Catalysis Reviews, 25*(2), 141–227.
17. Hutson, N. D., & Attwood, B. C. (2008). High temperature adsorption of CO_2 on various hydrotalcite-like compounds. *Adsorption, 14*(6), 781–789.
18. Hufton, J. R., Mayorga, S., & Sircar, S. (1999). Sorption-enhanced reaction process for hydrogen production. *AIChE Journal, 45*(2), 248–256.
19. Ding, Y., & Alpay, E. (2000). Equilibria and kinetics of CO_2 adsorption on hydrotalcite adsorbent. *Chemical Engineering Science, 55*(17), 3461–3474.

20. Ram Reddy, M. K., Xu, Z. P., & Diniz da Costa, J. C. (2008). Influence of water on high-temperature CO_2 capture using layered double hydroxide derivatives. *Industrial and Engineering Chemistry Research, 47*(8), 2630–2635.
21. Machado, B. F., & Serp, P. (2012). Graphene-based materials for catalysis. *Catalysis Science & Technology, 2*(1), 54–75.
22. Reijers, H. T. J., Boon, J., Elzinga, G. D., Cobden, P. D., Haije, W. G., & van den Brink, R. W. (2009). Modeling study of the sorption-enhanced reaction process for CO_2 Capture. I. Model development and validation. *Industrial and Engineering Chemistry Research, 48*(15), 6966–6974.
23. Lee, K. B., Verdooren, A., Caram, H. S., & Sircar, S. (2007). Chemisorption of carbon dioxide on potassium-carbonate-promoted hydrotalcite. *Journal of Colloid and Interface Science, 308* (1), 30–39.
24. Oliveira, E. L. G., Grande, C. A., & Rodrigues, A. E. (2008). CO_2 sorption on hydrotalcite and alkali-modified (K and Cs) hydrotalcites at high temperatures. *Separation and Purification Technology, 62*(1), 137–147.
25. Soares, J., Casarin, G., José, H., Moreira, R. P. M., & Rodrigues, A. (2005). Experimental and theoretical analysis for the CO_2 adsorption on hydrotalcite. *Adsorption, 11*(1), 237–241.
26. Halabi, M. H., de Croon, M. H. J. M., van der Schaaf, J., Cobden, P. D., & Schouten, J. C. (2012). High capacity potassium-promoted hydrotalcite for CO_2 capture in H_2 production. *International Journal of Hydrogen Energy, 37*(5), 4516–4525.
27. Kinniburgh, D. G. (1986). General purpose adsorption isotherms. *Environmental Science and Technology, 20*(9), 895–904.
28. Lee, K. B., Beaver, M. G., Caram, H. S., & Sircar, S. (2007). Chemisorption of carbon dioxide on sodium oxide promoted alumina. *AIChE Journal, 53*(11), 2824–2831.

Chapter 6
Influence of Alkali Metals on Layered Double Hydroxides Supported on Graphene Oxide for CO_2 Adsorption

6.1 Introduction

The use of supporting materials for layered double oxides (LDOs) has been demonstrated to be an effective approach to improve their stability in temperature-swing CO_2 adsorption cycles. Simultaneously, the intrinsic adsorption capacity of LDOs has been observed to increase although this requires high support contents that lead to very diluted hybrids with low CO_2 capacities per volume of adsorbent [1–3]. Graphene oxide appears as an outstanding support for LDOs since very modest amounts need to be added to observe a marked improvement in their multicycle stability (Chap. 5). Due to this high mass efficiency, the density of the adsorbents is not significantly reduced and adequate volumetric capacities are obtained. However, the capacities of LDO-GO hybrids are still relatively low and therefore it is necessary to explore strategies to increase the amount of CO_2 adsorbed without compromising the enhancement in the stability gained by the use of GO.

The presence of alkali ions on metal oxide surfaces is known to increase their basicity and therefore many researchers have observed a promoting effect of sodium, potassium and caesium on the CO_2 capacity of LDOs [4–7]. Bulky monovalent cations are weaker Lewis acids than divalent (Mg^{2+}) and trivalent (Al^{3+}) cations and this leads to the higher basicity of alkali-doped hydrotalcites. Some studies suggest that the interaction between the alkali salts and the aluminium oxide centres in the LDO plays a key role in the formation of strong basic sites which are more active for CO_2 adsorption [5]. Alternatively, the gain in capacity observed has been attributed to a higher concentration of surface defects created by the presence of the univalent alkali ions [7].

In the literature, most reports on alkali promotion of LDOs have focused on the assessment of unsupported materials [5, 6, 8, 9]. Recently Meis et al. [7] studied the effect that Na and K (either residues left in the washing step or deliberately impregnated) have on the adsorption capacity of LDOs supported on carbon

D. Iruretagoyena Ferrer, *Supported Layered Double Hydroxides as CO_2 Adsorbents for Sorption-enhanced H_2 Production*, Springer Theses,
DOI 10.1007/978-3-319-41276-4_6

nanofibers. A significant increase in the CO_2 uptake was observed in all cases although the capacities were slightly higher for the samples with alkali residues. The alkali metals seemed to have no effect on the stability of the hybrids. However, the authors only reported 10 adsorption-desorption cycles and the adsorption step was carried out in the presence of water which is known to slow down the rate of deactivation of LDOs [8]. Furthermore, very large carbon nanofibers contents (90 wt%) were used in the study and therefore it is not justifiable to extrapolate this observation to lower support loadings which are more relevant for industrial applications.

In this chapter the effect of alkali metals on the CO_2 adsorption performance of unsupported LDOs and LDO/GO hybrids is investigated in detail. A relatively low loading of GO (5 wt%) was selected for this study since an adequate stability was observed without compromising the CO_2 capacity per total volume of adsorbent (Sect. 5.4, Chap. 5). The materials were promoted either by incipient wetness impregnation or by leaving residual ions from the synthesis. Alkaline impurities are known to influence the CO_2 adsorption properties of the materials and therefore special attention is given to the assessment of the stability of the promoted adsorbents over multiple adsorption-desorption cycles under dry conditions. The chapter begins (Sect. 6.2) with a description of the methodology used to synthesise sodium doped Mg/Al LDHs and LDH/GO hybrids (5 wt% GO). Subsequently, Sect. 6.3 examines the physicochemical properties of the as-synthesised and activated adsorbents by a range of characterisation techniques including ICP, STEM, XRD, TGA and BET. CO_2-TPD was the main technique used to elucidate the nature of alkali metals promotion on LDH derivatives. Section 6.4 focuses on the CO_2 adsorption capacity and multicycle stability of the sodium promoted adsorbents while Sect. 6.5 consists in a comparative study of the promoting effect of sodium, potassium and caesium on unsupported and GO supported LDOs. In these two latter sections, the adsorption thermodynamics (isotherms) for relevant samples are presented and discussed. Finally, a summary of the key findings of the chapter is given in Sect. 6.6.

6.2 Synthesis of Sodium Promoted LDHs and LDH/GO Hybrids

Mg–Al unsupported LDHs and LDH/GO hybrids with nominal weight ratios 20/1 (LDH_{20}) were prepared via co-precipitation as described in Sects. 5.2.1 and 5.2.2 respectively (Chap. 5). The adsorbents were alkali promoted by two different methods. (a) The as-synthesised adsorbents were Na promoted by incipient wetness impregnation with an aqueous solution of Na_2CO_3. Subsequently, the materials were dried for 24 h at 393 K in a recirculating oven. The samples are denoted as $LDH_{I\text{-}Na}$ and $LDH_{20\text{-}I\text{-}Na}$. (b) After co-precipitation, the resulting suspension was aged and then filtered with minimum washing ($\sim$100 mL DI water) to leave residual sodium. The samples are denoted as $LDH_{R\text{-}Na}$ and $LDH_{20\text{-}R\text{-}Na}$.

Prior to the CO_2 adsorption measurements, all the adsorbents were activated in situ (in the TGA) at 673 K with N_2 to produce the corresponding unsupported LDOs (LDO, $LDO_{I\text{-}Na}$ and $LDO_{R\text{-}Na}$) and the LDO/GO hybrids (LDO_{20}, $LDO_{20\text{-}I\text{-}Na}$, $LDO_{20\text{-}R\text{-}Na}$). Further details of the activation procedure used are given in Sect. 5.2.3.

6.3 Characterisation

The unpromoted and sodium promoted adsorbents were characterised by inductively coupled plasma-optical emission spectroscopy (ICP-OES), X-ray diffraction (XRD), scanning transmission electron microscopy (STEM), thermogravimetric analysis (TGA), N_2 physisorption at 77 K and temperature-programmed desorption of CO_2 (CO_2-TPD). The detailed description of the equipment and the test procedures can be found in Chap. 3, (Sect. 3.4).

6.3.1 Composition of the Adsorbents

The compositions of the adsorbents are given in Table 6.1. The content of Mg, Al and Na in the materials was measured by ICP. The Mg/Al ratio of all the samples is close to the intended ratio of 2. Very low levels of residual sodium (less than 0.1 wt%) were found in the unpromoted adsorbents whereas the minimally washed and impregnated samples show sodium weight percentages between 1.7 and 2.8 wt%. In situ STEM-EDS analysis confirmed that magnesium, aluminium and sodium are distributed evenly in the materials, Fig. 6.1.

The TGA patterns of the LDH and hybrid samples are shown in Fig. 6.2. All the profiles exhibited three distinct regions of weight loss which can be attributed mainly to the dehydration, dehydroxylation and decarbonation of the materials. The actual GO content in the supported samples was determined from their corresponding TGA profiles as described in Sect. 5.3.1. As anticipated, the actual and nominal GO weight percentages in the hybrids were found to be very similar, Table 6.1. The loadings of GO presented in this study are low and therefore the loss of oxidative debris during the filtration step is expected to be very small.

6.3.2 Crystal Morphology

The crystallinity of the adsorbents was investigated by XRD, Fig. 6.3. All the samples show the typical reflections of hydrotalcite-like compounds [10]. No diffraction peaks associated with GO were observed for the LDH/GO hybrids due to the low content of carbon in the adsorbents. The diffraction patterns and crystallite

Table 6.1 Composition of the as-synthesised adsorbents

Sample name[a]	wt% GO Nom.	wt% GO actual	Mg/Al (mol/mol)	wt% Na Nom.	wt% Na actual
Pure LDH	0	0	2.2	0	0.02
$LDH_{I\text{-}Na}$	0	0	2.2	2	1.90
$LDH_{R\text{-}Na}$	0	0	2.2	–	2.80
LDH_{20}	5	4	2.2	0	0.07
$LDH_{20\text{-}I\text{-}Na}$	5	4	2.2	2	1.70
$LDH_{20\text{-}R\text{-}Na}$	5	6	2.2	–	1.90

[a]LDH_x, where x is the nominal LDH to GO weight ratio of the sample; I = impregnated, R = residual

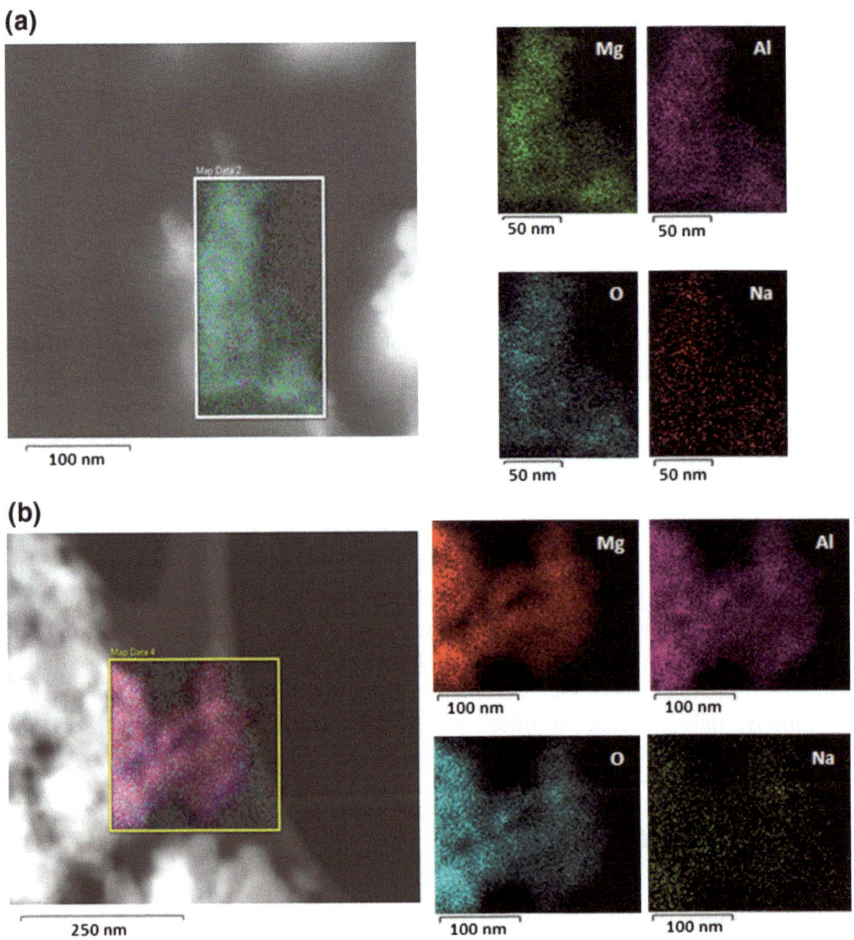

Fig. 6.1 STEM-EDS images of **a** $LDH_{I\text{-}Na}$ and **b** $LDH_{R\text{-}Na}$

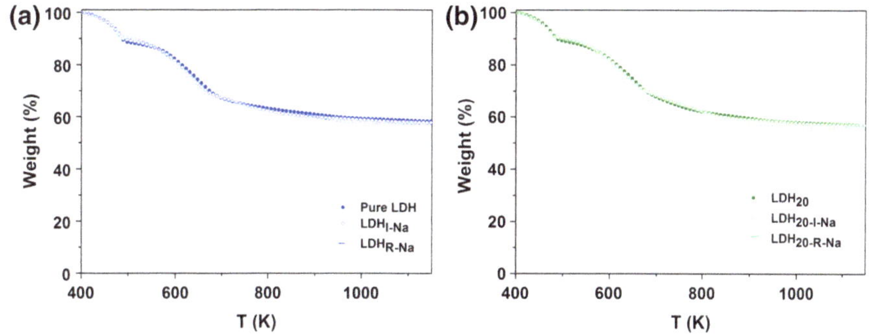

Fig. 6.2 Thermograms of **a** pure LDH and sodium promoted LDHs and **b** LDH_{20} and sodium promoted LDH_{20}

sizes of the samples impregnated with Na_2CO_3 ($LDH_{I\text{-}Na}$ and $LDH_{20\text{-}I\text{-}Na}$) are similar to those of the sodium free materials (LDH and LDH_{20}), Fig. 6.3 and Table 6.2. Therefore, it is likely that sodium is distributed homogeneously over the surface of the adsorbents. On the contrary, the crystallite sizes of the samples with residual sodium ($LDH_{R\text{-}Na}$ and $LDH_{20\text{-}R\text{-}Na}$) are larger than those of the thoroughly washed samples. This suggests that extensive washing during the filtration step decreases the final crystallite size of the material, probably due to erosion. For the minimum-washed samples, an additional diffraction line identified as $NaNO_3$ (104) is observed at 29.4°, exhibiting higher intensity for the unsupported LDH [11]. The intensity of this peak is related to the amount of crystalline sodium nitrate remaining in the sample. However, not all of the residual sodium may be in the form of crystalline agglomerates and part may be dispersed over the LDH. The ICP results for the thoroughly washed hybrids show that, although in very low amounts, some sodium ions are retained during the synthesis possibly due to the interaction with GO.

Fig. 6.3 XRD diffraction patterns of as-synthesised unpromoted and sodium promoted adsorbents. Characteristic reflections of LDH (♦) and $NaNO_3$ (❍)

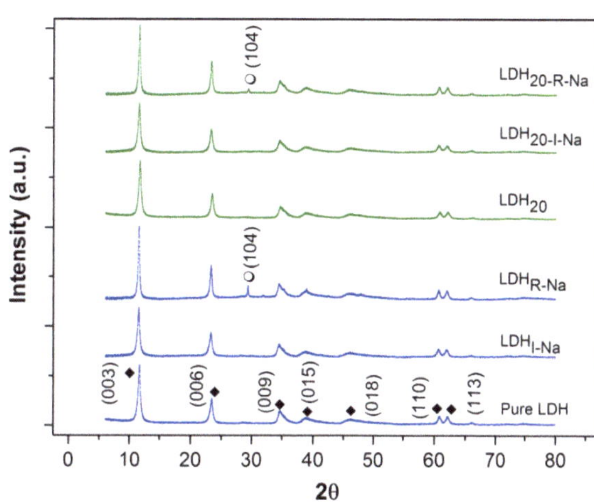

Table 6.2 Crystallite size of the as-synthesised and activated adsorbents

Sample name[a]	Crystallite size (003) (nm)	Sample name[a]	Crystallite size (200) (nm)
Pure LDH	21	Pure LDO	3.2
$LDH_{I\text{-}Na}$	18	$LDO_{I\text{-}Na}$	2.8
$LDH_{R\text{-}Na}$	29	$LDO_{R\text{-}Na}$	3.9
LDH_{20}	19	LDO_{20}	3.1
$LDH_{20\text{-}I\text{-}Na}$	18	$LDO_{20\text{-}I\text{-}Na}$	2.7
$LDH_{20\text{-}R\text{-}Na}$	24	$LDO_{20\text{-}R\text{-}Na}$	3.3

[a] LDH_x and LDO_x where x is the nominal LDH to GO weight ratio of the sample; I = impregnated, R = residual

TEM images of the pure LDH and LDH_{20} hybrid show small ($\sim$20 nm), crystalline, hexagonal platelets of LDH, Fig. 5.6 (Chap. 5). In agreement with the XRD results, the average LDH crystallite size of the sodium impregnated materials was similar to that of the unpromoted adsorbents whereas larger crystallites were observed for the minimum washed, especially in the case of the unsupported LDH, Fig. 6.4. It is worth mentioning that no GO nanosheets were visible in the TEM images of the Na free and Na promoted LDH/GO hybrids and therefore it is assumed that for these low loadings the GO nanosheets are well integrated with the LDH particles.

After calcination at 673 K the layered structure of the as-synthesised adsorbents collapses resulting in amorphous phases as revealed by XRD, Fig. 6.5. The diffractograms of the mixed oxides show broad peak reflections at $\sim$35°, $\sim$43° and $\sim$63° corresponding to the (111), (200) and (220) planes of periclase (MgO) [11]. Similar to the as-synthesised materials, the crystallite sizes of the samples with residual sodium were found to be slightly larger than those of the Na

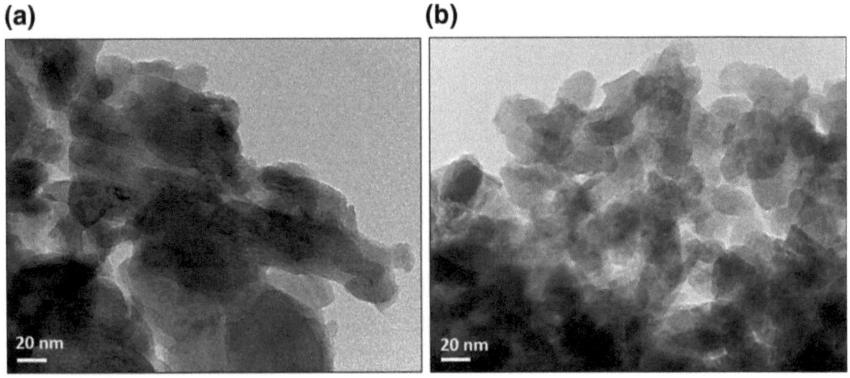

Fig. 6.4 Representative TEM images of $LDH_{R\text{-}Na}$ (**a**) and $LDH_{20\text{-}R\text{-}Na}$ (**b**)

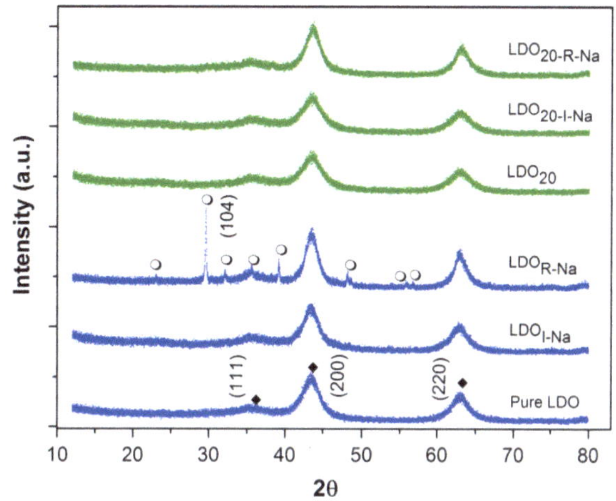

Fig. 6.5 XRD diffraction patterns of activated unpromoted and sodium promoted adsorbents. Characteristic reflections of LDO (♦) and $NaNO_3$ (O)

free and Na impregnated materials, Table 6.2. Additionally, no peaks associated with GO were detected for the activated hybrids.

The diffraction patterns of the activated samples impregnated with sodium ($LDO_{I\text{-}Na}$ and $LDO_{20\text{-}I\text{-}Na}$) present only the reflections of the pure LDO indicating that Na_2CO_3 remains homogeneously distributed after thermal treatment at 673 K. The XRD pattern of the hybrid with residual sodium ($LDO_{20\text{-}R\text{-}Na}$) also shows the diffraction peaks of the LDO exclusively while the unsupported LDO with residual sodium ($LDO_{R\text{-}Na}$) additionally exhibits the characteristic peaks of $NaNO_3$. The calcination temperature is above the melting point of sodium nitrate, which is 581 K, and therefore significant mobility of Na species is expected in the samples with residual sodium. It is likely that the GO present in $LDO_{20\text{-}R\text{-}Na}$ suppresses the agglomeration of sodium nitrate crystallites. It is also important to note that the $NaNO_3$ reflections observed in the XRD pattern of $LDO_{R\text{-}Na}$ are slightly shifted towards lower 2θ angles which may be caused by the partial decomposition of $NaNO_3$ into $NaNO_2$ and O_2 [12]. In fact, the CO_2-TPD of the unsupported LDO with residual sodium shows an oxygen peak with a maximum at 950 K. In contrast, no oxygen evolution is detected in the TPD profile of the hybrid containing residual sodium (see below). This may be related to the low concentration of $NaNO_3$ crystallites in the sample and to the fast consumption of any released oxygen in the combustion of GO.

6.3.3 *Textural Properties*

The textural properties of the adsorbents are given in Table 6.3. As expected, the surface area and mesopore volume of the materials increased considerably after

Table 6.3 Textural properties of as-synthesised and activated adsorbents

Sample name[a]	As-synthesised adsorbents			Sample name[a]	Activated adsorbents		
	SBET ($m^2\ g^{-1}$)	Vpore ($cm^3\ g^{-1}$)	dpore (nm)		SBET ($m^2\ g^{-1}$)	Vpore ($cm^3\ g^{-1}$)	dpore (nm)
Pure LDH	112	0.46	12	Pure LDO	223	0.51	10
$LDH_{I\text{-}Na}$	75	0.39	17	$LDO_{I\text{-}Na}$	90	0.45	16
$LDH_{R\text{-}Na}$	53	0.24	25	$LDO_{R\text{-}Na}$	73	0.31	24
LDH_{20}	106	0.54	15	LDO_{20}	199	0.62	13
$LDH_{20\text{-}I\text{-}Na}$	76	0.37	16	$LDO_{20\text{-}I\text{-}Na}$	102	0.43	14
$LDH_{20\text{-}R\text{-}Na}$	70	0.38	18	$LDO_{20\text{-}R\text{-}Na}$	99	0.40	14

[a]LDH_x and LDO_x, where x is the nominal LDH to GO weight ratio of the sample; I = impregnated, R = residual

thermal treatment at 673 K. All the as-synthesised and activated samples show a combination of type IV isotherms with H3 hysteresis loops indicating a wide pore size distribution, Fig. 6.6 [13]. The surface areas and pore volumes of the sodium promoted adsorbents were found to be significantly lower than those of the corresponding unpromoted materials. This decrease is possibly related to blockage/filling of the pores with sodium species. In addition, the reduction in surface area seems to be associated with the content of sodium in the sample. A decrease in surface area of ~33 % is observed for the adsorbents impregnated with sodium and the hybrid with residual sodium whereas the surface area of the minimum-washed LDH, which has ~1 wt% more sodium, is ~53 % lower than that of the pure LDH.

6.3.4 *Temperature-Programmed Desorption of CO_2*

The CO_2-TPDs of the unsupported LDOs (LDO, $LDO_{I\text{-}Na}$ and $LDO_{R\text{-}Na}$) and LDO/GO hybrids (LDO_{20}, $LDO_{20\text{-}I\text{-}Na}$ and $LDO_{20\text{-}R\text{-}Na}$) are presented in Fig. 6.7. All the TPD profiles show four desorption peaks corresponding to the formation of bicarbonates (α, 423 K), bidentate carbonates (β, 543 K) and monodentate carbonates (γ, 813 and δ, 963 K) [14]. As discussed in Chap. 5, the basic site density and distribution of the LDO derivatives are not affected by the presence of low GO contents (5 wt% GO, nominal). On the other hand, it is clear that relatively small amounts of sodium can alter considerably the basicity of the materials (Fig. 6.7, Tables 6.4 and 6.5). The release of bicarbonates formed on Bronsted OH^- groups (low temperature) is maintained whereas the adsorption in the intermediate temperature state corresponding to bidentate carbonates bonded to metal-oxygen pairs decreases. In addition, the number of monodentate carbonates adsorbed on low-coordination oxygen anions (high temperature) is significantly higher for the

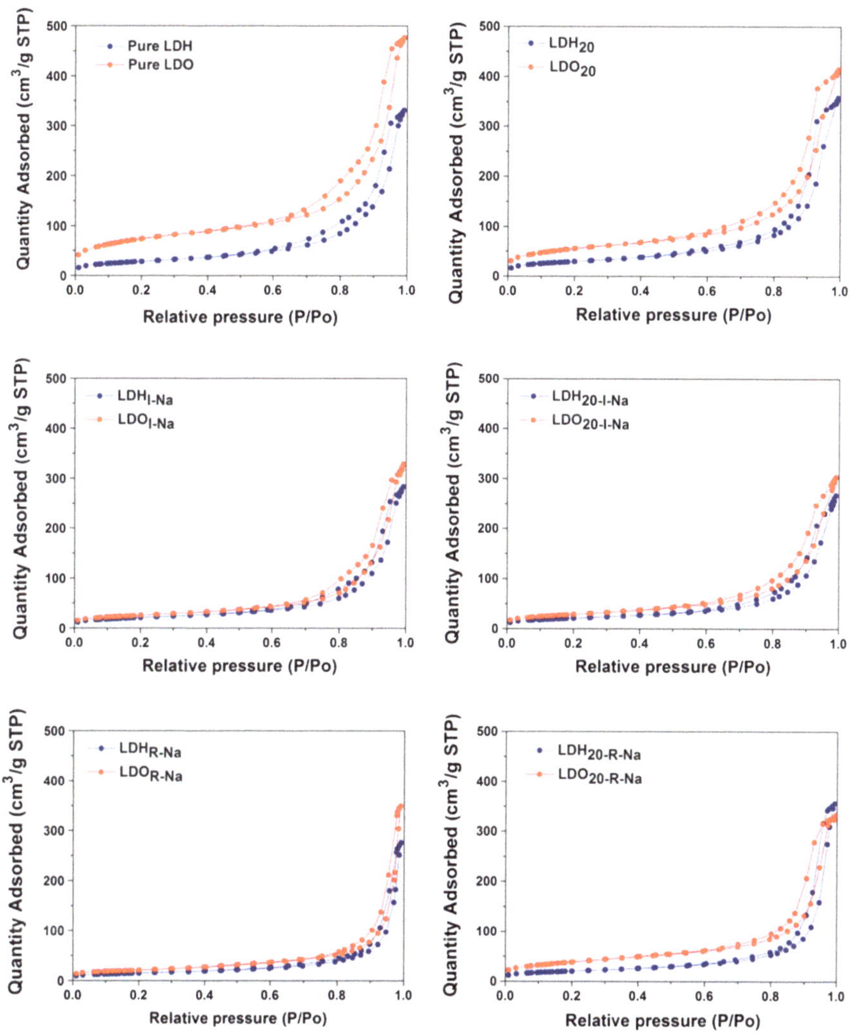

Fig. 6.6 N_2 physisorption isotherms of as-synthesised and activated adsorbents

sodium promoted samples, Table 6.5. The very strong basic sites that desorb CO_2 at 1115 K are possibly related to the decomposition of bulk sodium carbonates at the surface of the mixed oxides [5]. Overall, promotion by 2 % sodium approximately doubles the surface density of basic sites available after pretreatment at 673 K, Table 6.4. In passing, it is worth noting that the subtle differences in the TPD profiles of the alkali promoted samples can be attributed to the more homogeneous surface dispersion obtained by impregnation compared to the partial occlusion of sodium in the crystallites remaining in the minimum-washed samples.

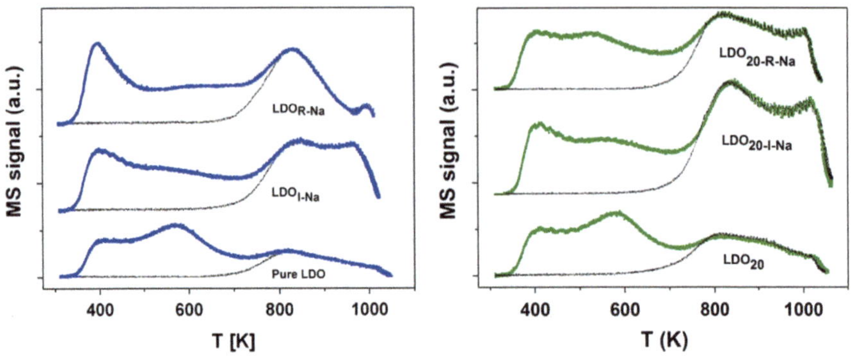

Fig. 6.7 CO_2-TPDs of LDOs (*right*) and LDO_{20}s (*left*) adsorbents following pretreatment at 673 K and adsorption at 313 K (*dark lines*). The blanks were subjected to pretreatment only (*light lines*)

Table 6.4 Density of basic sites: CO_2-TPD

Sample name[a]	wt% GO actual	Total CO_2 evolved (μmol g^{-1})	CO_2 evolved[b] (minus blank) (μmol g^{-1})	CO_2 evolved[b] (minus blank) (μmol g LDO^{-1})	CO_2 evolved[b] (minus blank) (μmol m^{-2})
Pure LDO	0	600	440	440	1.97
$LDO_{I\text{-}Na}$	0	910	450	459	5.00
$LDO_{R\text{-}Na}$	0	660	430	442	5.90
LDO_{20}	4	690	410	430	2.06
$LDO_{20\text{-}I\text{-}Na}$	4	1230	470	498	4.61
$LDO_{20\text{-}R\text{-}Na}$	6	920	450	489	4.55

[a]LDO_x, where x is the nominal LDO to GO weight ratio of the sample; I = impregnated, R = residual
[b]Values obtained by subtracting the corresponding blank from the total CO_2 evolved

Table 6.5 Percentage contribution of each desorption peak: CO_2-TPD (the deconvolution of the TPD profiles was carried out using Fityk 0.8.9)

Sample[a]	α (423 K)	β (543 K)	γ (813 K)	δ (963 K)
Pure LDO	13	55	28	4
$LDO_{I\text{-}Na}$	12	34	46	8
$LDO_{R\text{-}Na}$	19	44	36	1
LDO_{20}	12	49	33	6
$LDO_{20\text{-}I\text{-}Na}$	9	29	53	9
$LDO_{20\text{-}R\text{-}Na}$	8	34	53	5

[a]LDO_x, where x is the nominal LDO to GO weight ratio of the sample; I = impregnated, R = residual
[b]Bicarbonates (α), bidentates (β), monodentates (γ and δ)

6.4 CO_2 Adsorption Capacity and Multicycle Stability

CO_2 adsorption measurements were carried out using a TGA analyser (Perkin Elmer, TGA4000). The adsorbents were calcined in situ at 673 K under flowing N_2 for 4 h and then they were exposed to a CO_2 partial pressure of 200 mbar at 573 K. In temperature-swing cycling, the adsorption step was performed at 573 K for 1 h flowing the premixed CO_2 gas and the regeneration step was carried out at 673 K for 30 min flowing nitrogen. A flow rate of 20 mL min^{-1} was used during the tests. For further details the reader is referred to Chap. 3, Sect. 3.2.1.1.

Representative CO_2 adsorption profiles of the LDH derivatives are presented in Fig. 6.8. All the unpromoted and sodium promoted samples show similar adsorption kinetics exhibiting an initial fast stage followed by a slow uptake process. As shown in Chap. 5, the adsorption capacities per mass of LDO of the sodium free unsupported LDO and the LDO_{20} hybrid are the same within the experimental error, which is consistent with the TPD results presented above that revealed similar densities of basic sites for both adsorbents. Conversely, the first exposure capacities of the pure LDO and LDO_{20} hybrid were significantly increased by sodium promotion, Table 6.6. Since the surface area of the promoted adsorbents is lower than that of the sodium-free samples, the enhancement is attributable to the presence of additional interactions generated by the sodium cations.

As discussed above, the TPD profiles in Fig. 6.7 show that the presence of sodium increases the number of basic sites corresponding to the high temperature state. After activation, part of these stronger sites would be available for adsorption at 573 K and therefore it is likely that they are responsible for the observed increase in the adsorption capacity at the conditions used. These findings are consistent with other studies which have found that the presence of alkali ions on LDHs generates stronger basic sites with increased activity for CO_2 adsorption and base-catalysed reactions [5, 7, 15, 16]. Meis et al. suggested that the stronger basicity is related to

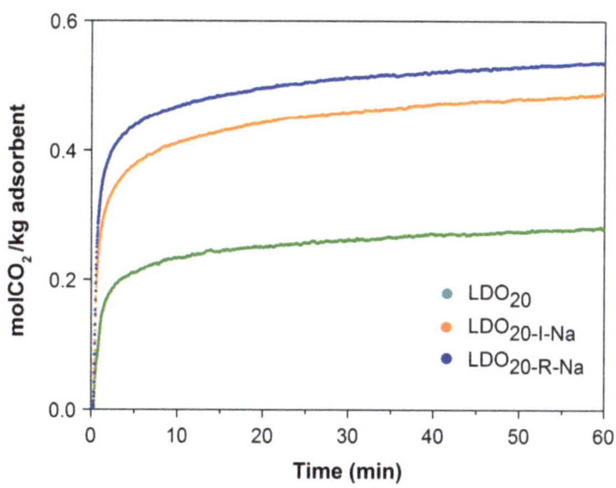

Fig. 6.8 Representative adsorption profiles of LDO_{20}, $LDO_{20\text{-I-Na}}$ and $LDO_{20\text{-R-Na}}$, at 573 K and PCO_2 = 200 mbar

Table 6.6 First-contact adsorption capacities of the adsorbents from TGA at 573 K and PCO_2 = 200 mbar

Sample name[a]	wt% GO actual	mol CO_2 kg LDO^{-1b}	mol CO_2 kg ads^{-1b}
Pure LDO	0	0.29 ± 0.01	0.29 ± 0.01
$LDO_{I\text{-}Na}$	0	0.56 ± 0.01	0.55 ± 0.01
$LDO_{R\text{-}Na}$	0	0.56 ± 0.02	0.54 ± 0.02
LDO_{20}	4	0.29 ± 0.03	0.28 ± 0.03
$LDO_{20\text{-}I\text{-}Na}$	4	0.52 ± 0.01	0.49 ± 0.01
$LDO_{20\text{-}R\text{-}Na}$	6	0.59 ± 0.02	0.54 ± 0.02

[a]LDO_x, where x is the nominal LDO to GO weight ratio of the sample; I = impregnated, R = residual

[b]The adsorption capacities are shown with standard errors

defects introduced in the LDO surface by the presence of alkali ions [7]. Additionally, Walspurger et al. found that the interaction between the aluminium centres in the mixed oxides and the alkali ions plays a crucial role in the formation of strongly basic sites which reversibly adsorb CO_2 at 573–773 K [5]. Other studies have found relatively low heats of adsorption for potassium promoted LDOs under dry conditions, indicating that physisorption or very weak-chemisorption are important in the CO_2 adsorption properties of LDOs at temperatures between 573 and 753 K [8, 17].

The adsorption capacity of the unsupported LDO increased by ~0.25 mol CO_2 kg^{-1} LDO irrespective of the method of alkali addition and within the experimental error. Increases of similar magnitude have been observed by other authors for significantly higher sodium contents [7] which may point to the existence of an optimum amount of alkali probably due to the occlusion of pores by Na_2CO_3 agglomerates [4]. For the LDO_{20} hybrid, the increase in capacity is slightly higher for the sample with residual sodium compared to the one impregnated with the same amount of promoter. This indicates that the impregnated Na_2CO_3 does not interact as efficiently as the residual sodium species. Analogous results have been reported for other carbon supports (carbon nanofibers) and have been related to a strong interaction between the alkali ions and the oxygen groups on the support during the impregnation which prevents an optimum incorporation of the promoter on the LDO [7]. In fact, a blank experiment using a GO sample loaded with 6 wt% Na (added as Na_2CO_3) showed an enhancement of capacity of 0.06 mol CO_2 kg^{-1}GO. This low promotion of CO_2 adsorption by sodium on the support demonstrates the necessity of having a good interaction between the alkali promoter and the LDO.

The profiles obtained from temperature-swing adsorption tests for the Na free and Na promoted adsorbents are presented in Fig. 6.9. The adsorption capacity of all the samples drops steeply during the first cycles and then becomes practically steady. Alpay et al. [8], Hutson et al. [17] and Hufton et al. [18] observed similar

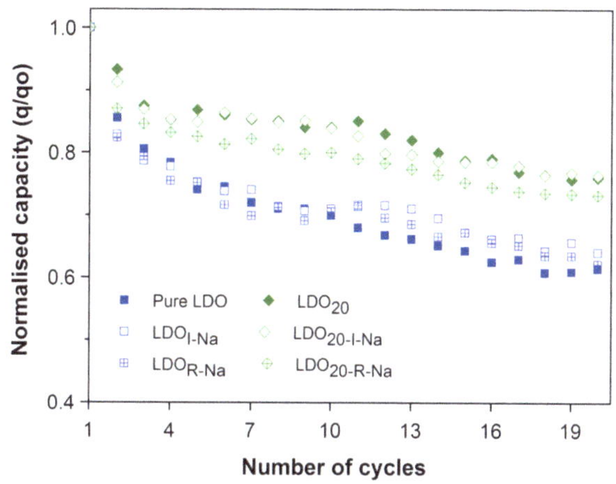

Fig. 6.9 Normalised CO_2 adsorption capacity over 20 adsorption-desorption cycles (adsorption 523 K; desorption 673 K)

deactivation trends for unpromoted and potassium promoted LDOs exposed to dry mixtures of CO_2. The authors attributed the initial loss in capacity to irreversible chemisorption of CO_2 on the fresh material. However, in the present study there is no evidence to indicate that irreversible CO_2 chemisorption is the main cause of deactivation. The fact that the initial drop in capacity depends on the content of GO suggests that rapid thermal sintering contributes significantly to the trend observed. The residual adsorption capacity after 20 cycles for the pure LDO is 75 % compared to 60 % for the LDO_{20} hybrid, in agreement with the multicycle tests discussed in Chap. 5. The alkali promoted adsorbents show TSA profiles very close to those of the corresponding sodium-free samples. This indicates that relatively low contents of sodium have no significant effect on the stability of the unsupported and the supported LDOs but enhance greatly their adsorption capacity.

As in the case of LDO_{20} (Chap. 5), the first-contact adsorption isotherm of the $LDO_{20\text{-R-Na}}$ hybrid was found to be better described by the Freundlich model whereas the equilibrium data after multicycle experiments fits the Langmuir equation, Fig. 6.10.[1] These findings are consistent with the similar thermal stability observed for the alkali free and the alkali loaded materials. Just after activation, the surface of the hybrids is heterogeneous and becomes more uniform over consecutive cycling. The increment in the value of the gas-solid interaction parameters (Table 6.7) compared to the unpromoted hybrid (Sect. 5.5.3) reveals an increase in the basic strength of the adsorption sites with sodium, which agrees with the TPD results presented above.

[1]The corresponding fittings to the Toth isotherm are presented in Appendix E.2.

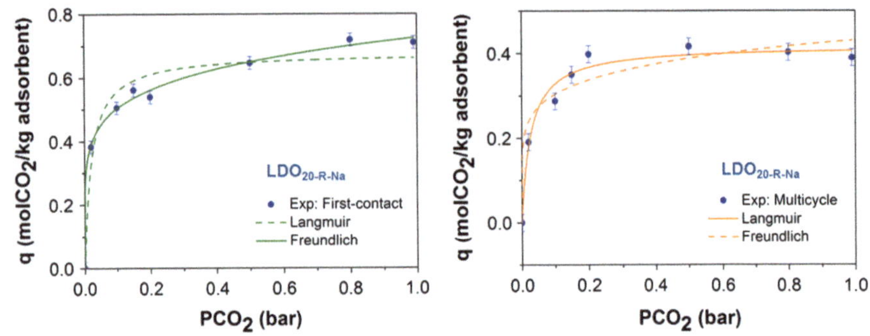

Fig. 6.10 First-contact and multicycle adsorption isotherm at 573 K under dry conditions from TGA measurements for $LDO_{20\text{-}R\text{-}Na}$; shown with standard errors

Table 6.7 Parameters for adsorption isotherms at 573 K from TGA under dry conditions

Adsorbent	Langmuir isotherm[a]			Freundlich isotherm[a]		
	m (mol kg^{-1})	b (1 bar^{-1})	R^2	k (mol kg^{-1})	n (–)	R^2
First-contact $LDO_{20\text{-}R\text{-}Na}$	0.68 ± 0.03	46.3 ± 15.3	0.95	0.73 ± 0.01	6.28 ± 0.39	0.99
Multicycle $LDO_{20\text{-}R\text{-}Na}$	0.42 ± 0.01	36.5 ± 8.5	0.98	0.43 ± 0.03	6.57 ± 1.68	0.93

[a]The parameters are shown with the standard errors obtained from the fittings

6.5 Influence of Na, K and Cs on LDHs Supported on GO for CO_2 Adsorption

The results discussed in the previous section indicate that the presence of sodium on pure and GO-supported LDOs enhances significantly their CO_2 capacity by increasing the density of strong adsorption sites. While it is expected that the extent of the promotion varies among the alkali metals, the studies reported in the literature hinder the identification of trends due to differences in the preparation methods and/or in the alkali loadings used [4, 6, 7, 15, 16]. The few comparative works available involve LDO samples with equal mass percentage of alkali, which results in marked differences in the number of promoting ions present on the surface of the materials [6, 15, 16]. In order to make a fair comparison of the influence of different alkali ions on the CO_2 adsorption capacity of LDO derivatives, the pure LDO and the LDO_{20} hybrid (5 wt% GO) were promoted with sodium, potassium or caesium seeking to add the same moles of alkali on all the samples. Alkali-containing materials were prepared by incipient impregnation from aqueous solutions of the corresponding carbonates. The nominal loadings were selected based on the Na_2CO_3 content of the samples presented in Sect. 6.3.1 and correspond to

Table 6.8 Composition of the as-synthesised adsorbents

Sample name[a]	wt% GO Nom.	wt% GO actual	Mg/Al (mol/mol)	mol alkali/kg LDH Nom.	mol alkali/kg LDH actual	wt% alkali actual
Pure LDH	0	0	2.3	0	0	0
$LDH_{I\text{-}Na}$	0	0	2.3	1.3	1.1	2.4
$LDH_{I\text{-}K}$	0	0	2.3	1.3	0.9	3.5
$LDH_{I\text{-}Cs}$	0	0	2.3	1.3	1.3	14.5
LDH_{20}	5	4.6	2.2	0	0	0
$LDH_{20\text{-}I\text{-}Na}$	5	4.6	2.2	1.4	1.1	2.5
$LDH_{20\text{-}I\text{-}K}$	5	4.6	2.2	1.4	0.9	3.3
$LDH_{20\text{-}I\text{-}Cs}$	5	4.6	2.2	1.4	1.3	14.4

[a]LDH_x, where x is the nominal LDO to GO weight ratio of the sample; I = impregnated

values slightly above the monolayer capacity.[2] In the literature most studies on alkali-promotion have dealt with loadings that are significantly above the monolayer capacity [6, 7]. The actual loadings of Na, K and Cs were measured by ICP-OES and ICP-AES/MS and were found to be very close to the intended values, Table 6.8.

The XRD diffraction patterns of the alkali promoted and unpromoted LDOs and LDO/GO hybrids are presented in Fig. 6.11. The as-synthesised adsorbents show the characteristic reflections of Mg–Al LDHs while the activated materials exhibit broad peaks corresponding to MgO. Similar to the homogeneous dispersion of alkali ions observed for the samples impregnated with sodium, no crystalline phases associated with the K and Cs salts were identified. The coherent sizes of the as-synthesised and activated samples calculated from the peaks (003) and (200) respectively were found to be slightly higher for the alkali promoted materials compared to the corresponding alkali free adsorbents (see Appendix E.1). These results differ from the hypothesis of Meis et al. [7] who proposed that the incorporation of Na and K ions in the surface of LDOs leads to a higher number of structural defects. However, it is worth mentioning that in contrast to the rest of the diffractograms presented in the thesis, the patterns shown in Fig. 6.11 were carried out in static mode due to problems with the spinning mechanism of the XRD instrument. This, added to the fact that the content of the LDH phase is lower due to dilution with the alkali promoter, contributes to the decreasing trend in the intensity of the LDH patterns observed (Cs < K < Na).

The distribution of K and Cs on the surface of the unsupported LDH was assessed by STEM-EDS, Fig. 6.12. Like the LDH sample impregnated with Na (see Fig. 6.1), the K and Cs atoms were found to be homogeneously distributed

[2]The monolayer capacity of the LDH based adsorbents correspond to 1.16, 0.79 and 0.51 mol alkali/kg LDH for Na, K and Cs respectively. These values were estimated using the van der Waals radius of the alkali ions and the surface area of the LDH (112 m^2/g).

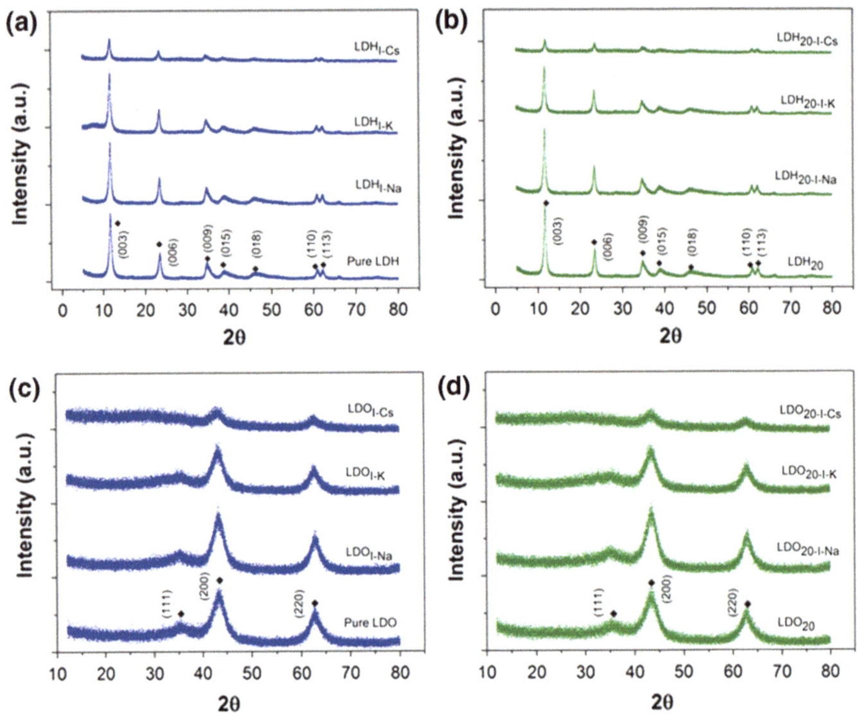

Fig. 6.11 XRD diffraction patterns of as-synthesised (**a** and **b**) and activated (**c** and **d**), unsupported (*blue*) and GO supported (*green*) adsorbents. Characteristic reflections of LDO (♦)

over the materials. However, it is observed that as the size of the alkali atom increases (i.e. Cs > K > Na), small aggregates of ions seem to be formed leaving areas of the surface uncovered. This is in agreement with the decrease in surface area observed in Table 6.9. It is likely that agglomerates of alkali ions block or partially fill pores of relatively small diameter.

To assess the effect of Na, K and Cs on the basic properties of the LDO phase, CO_2-TPD measurements were conducted using unsupported samples, Fig. 6.13. As expected, the presence of alkali ions over the surface of the adsorbents increases the number of active sites for CO_2 chemisorption. The distribution of basic strength was found to be similar for all the alkali doped materials (Table 6.11) whereas the total density of adsorption sites per mass of LDO increases in the order Cs > K > Na (i.e. following the order of basicity of the corresponding promoter oxide), Table 6.10. As the ionic radius of the alkali metals increases, the partial charge of the associated oxygen anions becomes more negative which results in the formation of sites of higher electron donor capacity [15]. This increases the basicity of the surface causing an increment in the CO_2 uptake of the materials. On the other hand, the tendency of Cs species to form bulky agglomerates results in a decrease in the number of accessible sites on the adsorbent and reduces the effectiveness of the

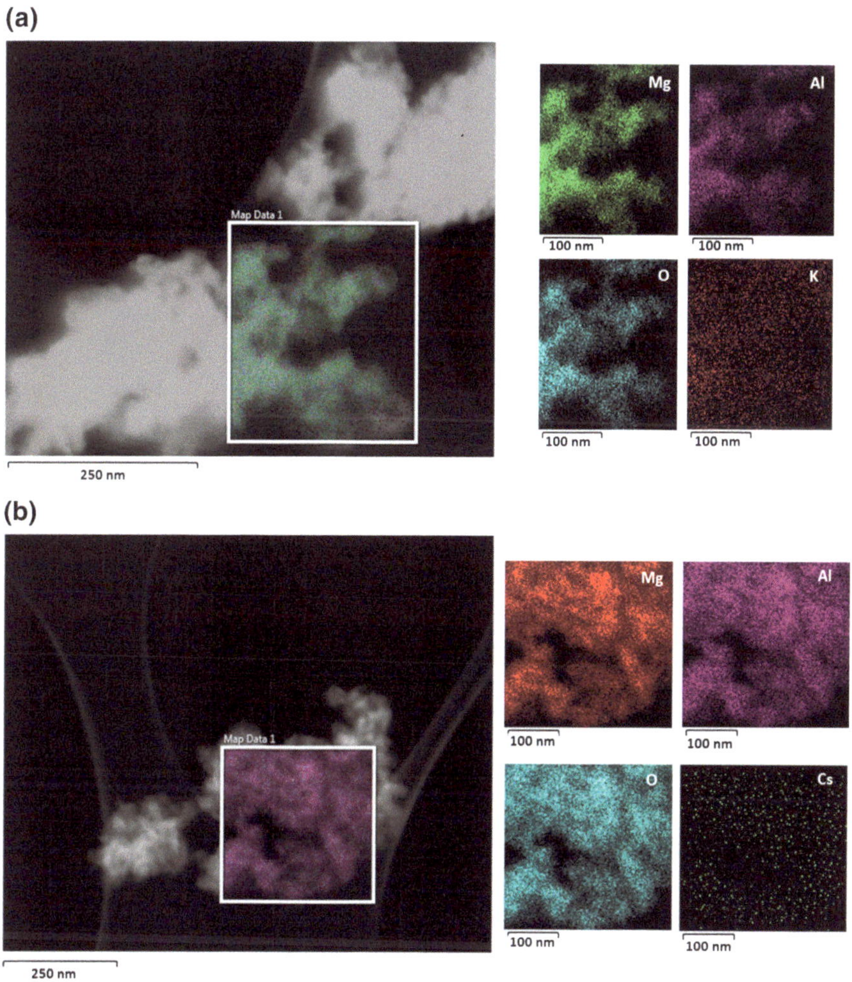

Fig. 6.12 STEM-EDS images of **a** $LDH_{I\text{-}K}$ and **b** $LDH_{I\text{-}Cs}$

interaction between the alkali and the LDO, hindering the intrinsic promotion of caesium. As a consequence, when the CO_2 capacities per mole of alkali are compared, there is a marked benefit in using potassium instead of sodium but a further increase in the basicity (and size) of the promoter to caesium is counter-productive. To disentangle these contradicting phenomena, it would be necessary to compare adsorbents with equal density of the alkali ions but in concentrations well below the monolayer capacity to procure an adequate (and comparable) incorporation of the promoters.

The adsorption capacity and multicycle stability of the alkali promoted LDOs and LDO/GO hybrids was determined by TGA analysis using the procedure described in Sect. 6.4. The capacities of LDO and LDO_{20} samples increased more

Table 6.9 Textural properties of as-synthesised and activated adsorbents

Sample name[a]	As-synthesised adsorbents				Sample name[a]	Activated adsorbents			
	SBET (m^2/g)	Vpore (cm^3/g)	dpore (nm)	µmol alkali/m^2		SBET (m^2/g)	Vpore (cm^3/g)	dpore (nm)	µmol alkali/m^2
Pure LDH	113	0.47	13	0	Pure LDO	226	0.54	9	0
$LDH_{I\text{-}Na}$	87	0.36	13	12.2	$LDO_{I\text{-}Na}$	116	0.45	12	9.1
$LDH_{I\text{-}K}$	87	0.38	13	10.1	$LDO_{I\text{-}K}$	98	0.41	14	9.0
$LDH_{I\text{-}Cs}$	64	0.28	14	17.1	$LDO_{I\text{-}Cs}$	67	0.28	14	16.3
LDH_{20}	108	0.50	14	0.0	LDO_{20}	215	0.55	10	0
$LDH_{20\text{-}I\text{-}Na}$	85	0.39	14	12.5	$LDO_{20\text{-}I\text{-}Na}$	116	0.47	14	9.2
$LDH_{20\text{-}I\text{-}K}$	79	0.37	15	10.5	$LDO_{20\text{-}I\text{-}K}$	100	0.43	14	8.4
$LDH_{20\text{-}I\text{-}Cs}$	67	0.34	16	16.1	$LDO_{20\text{-}I\text{-}Cs}$	84	0.36	15	12.9

[a]LDHx and LDOx, where x is the nominal LDH to GO weight ratio of the sample; I = impregnated

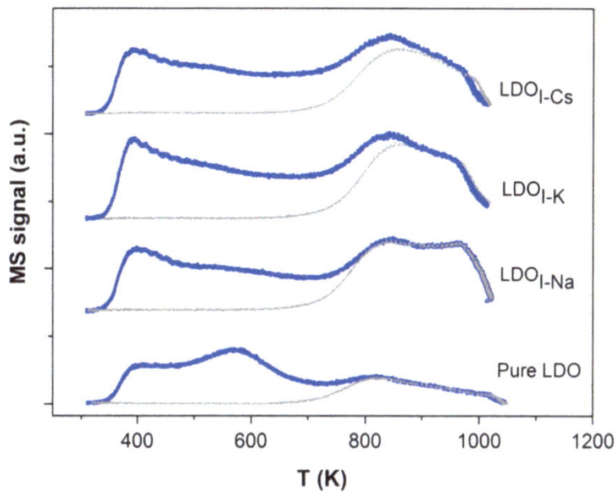

Fig. 6.13 CO_2-TPDs of pure LDO and alkali-LDOs following pretreatment at 673 K and adsorption at 313 K (*dark lines*). The blanks were subjected to pretreatment only (*light lines*)

Table 6.10 Density of basic sites: CO_2-TPD

Sample name[a]	Total CO_2 evolved (μmol g)	CO_2 evolved[b] (minus blank) (μmol g^{-1})	CO_2 evolved[b] (minus blank) (μmol g LDO^{-1})	CO_2 evolved[b] (minus blank) (μmol m^{-2})	mol CO_2 mol $alkali^{-1}$[c]
Pure LDO	599	437	437	1.93	0
LDO_{I-Na}	913	453	464	3.91	0.25
LDO_{I-K}	1023	616	639	6.29	0.43
LDO_{I-Cs}	919	554	648	8.27	0.30

[a] LDO_x, where x is the nominal LDO to GO weight ratio of the sample; I = impregnated
[b] Values obtained by subtracting the corresponding blank from the total CO_2 evolved
[c] Calculated using the mol alkali/kg LDO given in Table 6.13

Table 6.11 Percentage contribution of each desorption peak: CO_2-TPD (the deconvolution of the TPD profiles was carried out using Fityk 0.8.9)

Sample[a]	α (423 K)	β (543 K)	γ (813 K)	δ (963 K)
Pure LDO	13	55	28	4
LDO_{I-Na}	12	34	46	8
LDO_{I-K}	12	35	50	3
LDO_{I-Cs}	11	36	49	4

[a] LDO_x, where x is the nominal LDO to GO weight ratio of the sample; I = impregnated
[b] Bicarbonates (α), bidentates (β), monodentates (γ and δ)

Table 6.12 First-contact adsorption capacities of the adsorbents at 573 K and PCO_2 = 200 mbar

Sample name[a]	Alkali (wt%)	mol alkali/kg LDH[b]	mol CO_2/kg LDO[c]	mol CO_2/kg ads[c]	μmol CO_2/m^2	mol CO_2/mol alkali
Pure LDO	0	0	0.29 ± 0.01	0.29 ± 0.01	1.3	0
$LDO_{I\text{-}Na}$	2.4	1.1	0.57 ± 0.01	0.56 ± 0.01	4.8	0.31
$LDO_{I\text{-}K}$	3.5	0.9	0.71 ± 0.02	0.69 ± 0.02	7.0	0.47
$LDO_{I\text{-}Cs}$	14.5	1.3	0.81 ± 0.02	0.69 ± 0.02	10.4	0.37
LDO_{20}	0	0	0.29 ± 0.03	0.28 ± 0.03	1.3	0
$LDO_{20\text{-}I\text{-}Na}$	2.5	1.1	0.52 ± 0.02	0.49 ± 0.02	4.2	0.28
$LDO_{20\text{-}I\text{-}K}$	3.3	0.9	0.66 ± 0.02	0.61 ± 0.02	6.1	0.44
$LDO_{20\text{-}I\text{-}Cs}$	14.4	1.3	0.79 ± 0.02	0.64 ± 0.02	7.6	0.36

[a]LDO_x, where x is the nominal LDO to GO weight ratio of the sample; I = impregnated
[b]Converted to mol alkali/kg LDO considering that 60 % of the mass of the adsorbent is lost during activation
[c]The adsorption capacities are shown with standard errors

than 40 % with the addition of the alkali ions, Table 6.12. Under the operating conditions used, the capacity of the promoted samples per mass of LDO increases following the same trend observed in the CO_2-TPD tests, i.e. Cs > K > Na. The capacity per mole of alkali was found to be larger for K compared to Na and Cs, showing that there is no advantage in the use of caesium (which is a stronger base) as a promoter for the metal loadings used in the present study. The effectiveness of the alkali ions as promoters is greater for the unsupported samples compared to the GO containing adsorbents. It is likely that during impregnation the alkali ions interact with oxygen species from the support preventing an optimum incorporation of the promoter in the LDO. The comparison between the supported solids with impregnated and residual sodium in Sect. 6.4 leads to the same conclusion.

The adsorption capacity of the unsupported $LDO_{I\text{-}K}$ sample (0.69 mol CO_2/kg ads) is comparable to the CO_2 capacities reported by Ding et al. [8] (0.4 mol CO_2/kg ads) and Hufton et al. [18] (0.52 mol CO_2/kg ads) when differences in adsorption temperature and CO_2 partial pressure are taken into account. In contrast, higher intrinsic capacities were obtained by Meis et al. [7] for K-LDO/CNF (0.8 mol CO_2/kg LDO) and Na-LDO/CNF (2.2 mol CO_2/kg LDO) adsorbents. However, these measurements were carried out at a lower adsorption temperature and in the presence of water, which is known to increase the adsorption capacity of LDOs. In addition, the corresponding adsorption capacities per total mass of adsorbent are relatively low (0.08 and 0.22 mol CO_2/kg ads for K-LDO/CNF and Na-LDO/CNF respectively) since large amounts of support were used (90 wt% CNF). A better compromise between adsorption capacity (0.49 and 0.61 molCO_2/kg ads for $LDO_{20\text{-}I\text{-}Na}$ and $LDO_{20\text{-}I\text{-}K}$) and amount of support used (5 wt% GO) is achieved in the present study. Table 6.13 summarises the adsorption capacities and operating conditions of various alkali promoted LDOs from the literature.

Continuous adsorption-desorption cycles reveal that all the alkali LDO/GO samples are consistently more stable than the alkali unsupported adsorbents.

Table 6.13 Adsorption capacities of alkali promoted LDOs

Authors	Adsorbent	Support (wt%)	Alkali (wt%)	mol CO_2/kg LDO	mol CO_2/kg ads	Conditions
Present research	$LDO_{I\text{-}Na}$	0	2.4	0.57	0.56	Dry, 573 K, PCO_2 = 200 mbar
Present research	$LDO_{I\text{-}K}$	0	3.5	0.71	0.69	Dry, 573 K, CO_2 = 200 mbar
Alpay et al. [8]	$LDO_{\text{-}K}$	0	–	0.40[a]	0.40[a]	Dry, 673 K, PCO_2 = 100 mbar
Hufton et al. [18]	$LDO_{\text{-}K}$	0	–	0.52[a]	0.52[a]	Dry, 673 K, PCO_2 = 700 mbar
Oliveira et al. [6]	$LDO_{\text{-}Cs}$	0	20	0.44	0.44	Wet, 673 K, PCO_2 = 400 mbar
Oliveira et al. [6]	$LDO_{\text{-}K}$	0	20	0.76	0.76	Wet, 673 K, PCO_2 = 400 mbar
Present research	$LDO_{20\text{-}I\text{-}Na}$	GO (4.6 wt%)	2.5	0.52	0.49	Dry, 573 K, PCO_2 = 200 mbar
Present research	$LDO_{20\text{-}I\text{-}K}$	GO (4.6 wt%)	3.3	0.66	0.61	Dry, 573 K, PCO_2 = 200 mbar
Meis et al. [7]	LDO/CNF (R-Na)	CNF (90 wt%)	12	2.2	0.22	Wet, 523 K, PCO_2 = 50 mbar
Meis et al. [7]	LDO/CNF (I-K)	CNF (90 wt%)	11	0.8	0.08	Wet, 523 K, PCO_2 = 50 mbar

[a]The alkali loadings are not specified in the publications

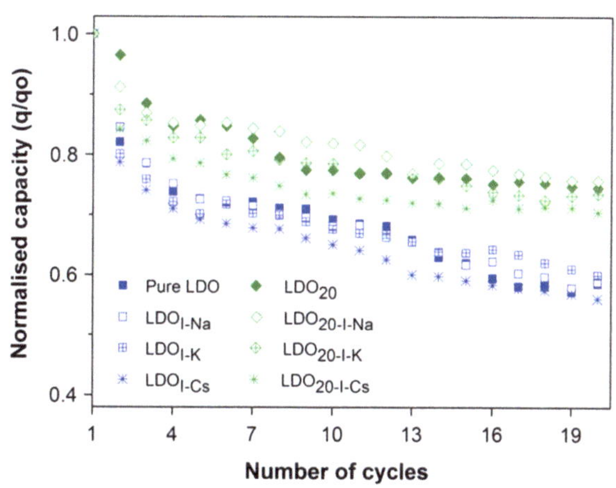

Fig. 6.14 Normalised CO_2 adsorption capacity over 20 adsorption-desorption cycles (adsorption 573 K; desorption 673 K)

Remarkably, as observed previously for the sodium containing materials, the amounts of potassium and caesium used in this study do not compromise the stability imparted by GO (Fig. 6.14).

As anticipated, the first-contact adsorption isotherms of the alkali promoted LDO and LDO_{20} hybrids were better described by the Freundlich model whereas due to

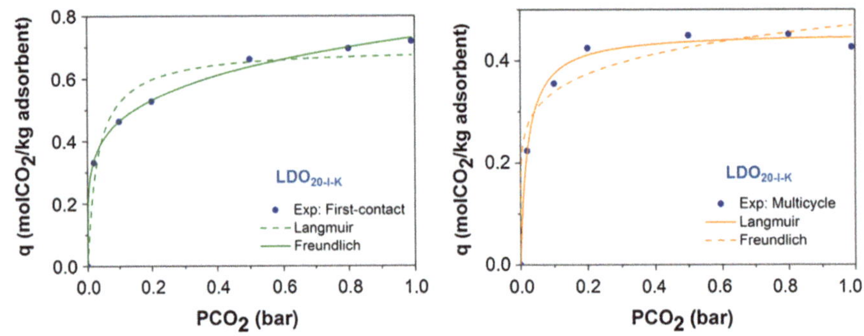

Fig. 6.15 First-contact and multicycle adsorption isotherm at 573 K under dry conditions from TGA measurements for $LDO_{20\text{-I-K}}$

Table 6.14 Parameters for adsorption isotherms at 573 K of potassium promoted samples from TGA under dry conditions

Adsorbent	Langmuir isotherm[a]			Freundlich isotherm[a]		
	m (mol/kg)	b (1/bar)	R^2	k (mol/kg)	n (–)	R^2
First-contact $LDO_{I\text{-}K}$	0.69 ± 0.03	38.7 ± 10	0.97	0.71 ± 0.01	5.8 ± 0.47	0.99
Multicycle $LDO_{I\text{-}K}$	0.49 ± 0.01	42.6 ± 6.7	0.99	0.50 ± 0.02	6.5 ± 1.2	0.96
First-contact $LDO_{20\text{-I-K}}$	0.70 ± 0.04	29.7 ± 9.8	0.95	0.73 ± 0.01	5.0 ± 0.21	0.99
Multicycle $LDO_{20\text{-I-K}}$	0.46 ± 0.01	46.1 ± 6.4	0.99	0.47 ± 0.02	7.1 ± 1.7	0.94

[a]The parameters are shown with the standard errors obtained from the fittings

the relatively low content of GO in the samples (5 wt% GO) the isotherms after 20 multiple cycles were found to fit better to the Langmuir equation, Fig. 6.15 and Table 6.14.[3] Hybrids with higher loadings of GO are expected to preserve the heterogeneity of the adsorption sites for a longer number of cycles as it was observed for an alkali-free hybrid containing 17 wt% GO (Sect. 5.5, Chap. 5).

As discussed in Chap. 2, most of the research concerning alkali based LDOs has focused on investigating the adsorption properties of potassium promoted LDOs [5, 8, 9]. It can be seen from Table 6.15 that the first-contact adsorption parameters of the K-LDO sample of the present study are in the same order of magnitude as those reported by Halabi et al. [9] and Reijers et al. [19]. In addition, although a better agreement is found using the Freundlich equation for the first-contact equilibrium data, the monolayer capacity and the interaction parameter obtained using the Langmuir model are comparable with first-contact data sets reported previously for potassium promoted LDOs, Table 6.15 [8, 18, 20]. The isosteric heats of

[3]The fitting parameters corresponding to multicycle adsorption isotherms at 573 K under dry conditions from TGA measurements for the Na and Cs promoted LDOs and LDO_{20} hybrids are shown in Appendix E.3.

Table 6.15 Parameters for adsorption isotherms from alkali promoted LDO at 573 K under dry conditions

Authors	Adsorbent	Langmuir isotherm[a]		Freundlich isotherm[a]	
		m (mol/kg)	b (1/bar)	k (mol/kg)	n (–)
Present work (first-contact)	$LDO_{I\text{-}K}$	0.69 ± 0.03	38.7 ± 10	0.71 ± 0.01	5.8 ± 0.47
Present work (first-contact)	$LDO_{20\text{-}I\text{-}K}$	0.7 ± 0.04	29.7 ± 9.8	0.73 ± 0.01	5.02 ± 0.21
Hufton et al. [18]	K promoted LDO	0.45	94.6[b]	–	–
Ding and Alpay et al. [8]	K promoted LDO	0.63	23.1[b]	–	–
Lee et al. [21]	K promoted LDO	0.25	73.3[b]	–	–
Halabi et al. [9]	K promoted LDO	–	–	1.04[c]	1.68[c]

[a]The parameters are shown with the standard errors obtained from the fittings
[b]The values of b at 573 K were obtained using the heats of adsorption reported in the references
[c]Parameters reported at 673 K

Table 6.16 Isosteric heats of adsorption

Authors	Adsorbent	First-contact	Multicycle
		Q_{ads} (kJ/mol)	Q_{ads} (kJ/mol)
Present work	$LDO_{20\text{-}I\text{-}K}$	20	22
Alpay et al. [8]	K promoted LDO	10	–
Lee et al. [20]	K promoted LDO	21	–
Hufton et al. [18]	K promoted LDO	29	–

adsorption, Q_{ads}, of the $LDO_{20\text{-}I\text{-}K}$ samples were obtained from the adsorption isotherms at 573 and 623 K at a constant uptake of 0.01 mol kg^{-1}. A number of studies have found similar Henry's law isosteric heats under dry conditions for potassium promoted LDOs (Table 6.16) and have attributed the relatively low values obtained to a physisorption phenomenon or to severe chemical alteration of the adsorbent surface.

6.6 Concluding Remarks

In this chapter the influence of alkali metals (Na, K and Cs) on the CO_2 adsorption properties of unsupported and GO supported LDOs (5 wt%) has been investigated in detail. It was found that while the stability of the adsorbents in temperature-swing multicycles is not influenced significantly by the promoter in the loadings studied ($\sim$1 mol alkali/kg LDH), the addition of alkali ions increases by more than 40 %

the amount of CO_2 retained by the materials at 573 K. The incorporation of alkali on the supported and the unsupported adsorbents results in a greater density of strong basic sites which correlates with the increase in the CO_2 capacity observed.

The distribution of basic strength was found to be similar for all the alkali doped materials but the enhancement in the density of chemisorption sites (CO_2-TPD) and in the total capacity of the adsorbents at 573 K varies among the different alkali metals. While the increase in the basicity of the promoter (Cs > K > Na) is beneficial for the CO_2 capacity at 573 K when Na and K are compared, the tendency of Cs species to form bulky agglomerates results in a decrease in the accessible surface of the adsorbent and hinders an effective interaction between Cs and the LDO. In addition, the Cs samples were found to be slightly less stable than the K and Na promoted samples, the latter two showing similar multicycle performances to those of the corresponding unpromoted materials. Consequently, for the alkali contents tested in this study, potassium appears as the most efficient promoter for CO_2 adsorption.

An investigation of the influence of the strategy followed to introduce sodium in the samples reveal that the promotion is more efficient when the ions are left as residues from the synthesis compared to the a posteriori addition by impregnation. The difference is more noticeable for the LDO/GO hybrids, where the incorporation of alkali in the LDO is more effective for the residual route thereby resulting in higher adsorption capacities. On the other hand, during the impregnation of Na from aqueous solutions, the alkali cations tend to interact strongly with the oxygen containing species from the carbon support, decreasing the actual loading of promoter in the LDO. In agreement with these results it was observed that the effectiveness of alkali (Na, K and Cs) promotion by impregnation is greater for the unsupported samples compared to the GO containing adsorbents.

The CO_2 adsorption capacities of the alkali promoted materials of this study are comparable to values reported in the literature under dry conditions. In addition, the isosteric heat of adsorption and the parameters obtained using the Langmuir and Freundlich models are of the same order of magnitude that data sets reported for potassium promoted LDOs.

References

1. Othman, M. R., Rasid, N. M., & Fernando, W. J. N. (2006). Mg–Al hydrotalcite coating on zeolites for improved carbon dioxide adsorption. *Chemical Engineering Science, 61*(5), 1555–1560.
2. Meis, N. N. A. H., Bitter, J. H., & de Jong, K. P. (2009). Support and size effects of activated hydrotalcites for precombustion CO_2 capture. *Industrial and Engineering Chemistry Research, 49*(3), 1229–1235.
3. Aschenbrenner, O., McGuire, P., Alsamaq, S., Wang, J., Supasitmongkol, S., Al-Duri, B., et al. (2011). Adsorption of carbon dioxide on hydrotalcite-like compounds of different compositions. *Chemical Engineering Research and Design, 89*(9), 1711–1721.
4. Yang, J.-I., & Kim, J.-N. (2006). Hydrotalcites for adsorption of CO_2 at high temperature. *Korean Journal of Chemical Engineering, 23*(1), 77–80.

5. Walspurger, S., Boels, L., Cobden, P. D., Elzinga, G. D., Haije, W. G., & van den Brink, R. W. (2008). The crucial role of the K+–Aluminium oxide interaction in K+−promoted alumina- and hydrotalcite-based materials for CO_2 sorption at high temperatures. *ChemSusChem, 1*(7), 643–650.
6. Oliveira, E. L. G., Grande, C. A., & Rodrigues, A. E. (2008). CO_2 sorption on hydrotalcite and alkali-modified (K and Cs) hydrotalcites at high temperatures. *Separation and Purification Technology, 62*(1), 137–147.
7. Meis, N. N. A. H., Bitter, J. H., & de Jong, K. P. (2010). On the influence and role of alkali metals on supported and unsupported activated hydrotalcites for CO_2 sorption. *Industrial and Engineering Chemistry Research, 49*(17), 8086–8093.
8. Ding, Y., & Alpay, E. (2000). Equilibria and kinetics of CO_2 adsorption on hydrotalcite adsorbent. *Chemical Engineering Science, 55*(17), 3461–3474.
9. Halabi, M. H., de Croon, M. H. J. M., van der Schaaf, J., Cobden, P. D., & Schouten, J. C. (2012). High capacity potassium-promoted hydrotalcite for CO_2 capture in H_2 production. *International Journal of Hydrogen Energy, 37*(5), 4516–4525.
10. Millange, F., Walton, R. I., & O'Hare, D. (2000). Time-resolved in situ X-ray diffraction study of the liquid-phase reconstruction of Mg–Al–carbonate hydrotalcite-like compounds. *Journal of Materials Chemistry, 10*(7), 1713–1720.
11. León, M., Díaz, E., Bennici, S., Vega, A., Ordóñez, S., & Auroux, A. (2010). Adsorption of CO_2 on hydrotalcite-derived mixed oxides: Sorption mechanisms and consequences for adsorption irreversibility. *Industrial and Engineering Chemistry Research, 49*(8), 3663–3671.
12. Freeman, E. S. (1956). The kinetics of the thermal decomposition of sodium nitrate and of the reaction between sodium nitrite and oxygen. *The Journal of Physical Chemistry, 60*(11), 1487–1493.
13. Rouquerol, F., Rouquerol, J., & Sing, K. (1999). *Adsorption by powders & porous solids*. London, UK: Academic Press.
14. Di Cosimo, J. I., Apesteguia, C. R., Ginés, M. J. L., & Iglesia, E. (2000). Structural requirements and reaction pathways in condensation reactions of alcohols on MgyAlOx catalysts. *Journal of Catalysis, 190*(2), 261–275.
15. Di Cosimo, J. I., Díez, V. K., & Apesteguía, C. R. (1996). Base catalysis for the synthesis of α, β-unsaturated ketones from the vapor-phase aldol condensation of acetone. *Applied Catalysis, A: General, 137*(1), 149–166.
16. Abelló, S., Medina, F., Tichit, D., Pérez-Ramírez, J., Rodríguez, X., Sueiras, J. E., et al. (2005). Study of alkaline-doping agents on the performance of reconstructed Mg–Al hydrotalcites in aldol condensations. *Applied Catalysis, A: General, 281*(1–2), 191–198.
17. Hutson, N. D., & Attwood, B. C. (2008). High temperature adsorption of CO_2 on various hydrotalcite-like compounds. *Adsorption, 14*(6), 781–789.
18. Hufton, J. R., Mayorga, S., & Sircar, S. (1999). Sorption-enhanced reaction process for hydrogen production. *AIChE Journal, 45*(2), 248–256.
19. Reijers, H. T. J., Boon, J., Elzinga, G. D., Cobden, P. D., Haije, W. G., & van den Brink, R. W. (2009). Modeling study of the sorption-enhanced reaction process for CO_2 Capture. I. Model development and validation. *Industrial and Engineering Chemistry Research, 48*(15), 6966–6974.
20. Lee, K. B., Beaver, M. G., Caram, H. S., & Sircar, S. (2007). Chemisorption of carbon dioxide on sodium oxide promoted alumina. *AIChE Journal, 53*(11), 2824–2831.
21. Lee, K. B., Verdooren, A., Caram, H. S., & Sircar, S. (2007). Chemisorption of carbon dioxide on potassium-carbonate-promoted hydrotalcite. *Journal of Colloid and Interface Science, 308*(1), 30–39.

Chapter 7
CO_2 Adsorption on Unsupported and Graphene Oxide Supported Layered Double Hydroxides in a Fixed-Bed

7.1 Introduction

The studies presented in the previous chapters have dealt with different strategies to improve the CO_2 capacity of LDO-based materials and accordingly, emphasis has been given to the measurement of adsorption equilibrium. However, in practical applications, adsorbents are usually contacted with gases flowing through packed beds where adequate uptake rates are as crucial as high equilibrium capacities. Consequently, the assessment of the overall dynamics of adsorption and the development of mathematical models to predict the separation performance is very important from the perspective of design engineering.

The information needed to evaluate the kinetics of adsorption can be obtained from the measurement of the response of the packed column to a positive step change in the CO_2 concentration (breakthrough curve). Typically there is a time delay in the response due to the gas hold-up of the column and to the adsorptive uptake, and a change in the slope of the breakthrough due to the combined effect of axial dispersion and resistance to transport. Information on the adsorption kinetics can be extracted by matching the experimental breakthrough profile with a response curve calculated from a suitable dynamic model.

In this chapter, the adsorption of CO_2 over unsupported and GO-supported LDO is studied in transient experiments conducted in a fixed-bed column. The objective is twofold: to procure information that allows the adsorption kinetics to be evaluated, and to measure the CO_2 adsorption capacity in the presence of water (which could not be measured by TGA). In Sect. 7.2 the experimental CO_2 breakthrough curves are presented. First the dry adsorption capacities determined by breakthrough analysis are compared to the corresponding TGA values to validate the methodology used. Subsequently the results for dry and wet CO_2 adsorption are discussed and the multicycle stability is evaluated in temperature-swing and

D. Iruretagoyena Ferrer, *Supported Layered Double Hydroxides as CO_2 Adsorbents for Sorption-enhanced H_2 Production*, Springer Theses,
DOI 10.1007/978-3-319-41276-4_7

isothermal N_2 purge experiments. Section 7.3 introduces the dynamic model used to analyse the kinetic performance of the adsorbents. The importance of mass transfer limitations in the adsorption of CO_2 on pure LDO and on a hybrid containing 5 wt% GO is assessed through a parameter estimation based on the linear driving force (LDF) approximation. Finally, a critique of the findings and a discussion on the limitations of the studies are given in Sect. 7.4.

7.2 Breakthrough Curve Experiments

CO_2 breakthrough measurements were conducted in a fixed-bed rig designed and built as part of this research. The adsorption column (d_t = 9.5 mm, ID = 7.7 mm, length = 364 mm) was divided into 3 sections: preheat, adsorbent and exit. The adsorption section was packed with 0.5–1 g of sorbent material (250–500 µm), which resulted in bed heights between 16.5 and 33.1 mm. The effluent of the column was monitored with a mass spectrometer. A schematic representation and a detailed description of the system are given in Sect. 3.2.2 (Chap. 3). A typical experimental run involved the following steps: (1) activation in situ with N_2 at 673 K; (2) feed gas switch to CO_2/Ar or CO_2/Ar/H_2O at 573 K to initiate the adsorption step; (3) regeneration by isothermal N_2 purge or by temperature-swing at 673 K under N_2. Table 7.1 summarises the standard operating conditions used.

The amount of CO_2 adsorbed was calculated using a graphical method and the material balance in the column (see Sect. 3.2.2). The adsorption capacities were obtained by dividing the moles of CO_2 adsorbed over the mass of solid recovered from the column (m_a). There was a ~40 % decrease in the mass of the recovered adsorbent with respect to the initial loading due to the dehydration and partial decarbonation of the materials during the activation step.

Table 7.1 Standard experimental conditions for adsorption measurements

	Total flow (mL min^{-1}) STP	Composition	Temperature (K)
Activation	138	100 % N_2	RT—673 K (10 K min^{-1}, 4 h)
Adsorption dry wet	46–138 55–165	2–99 % CO_2/Ar 1.7–42 % CO_2/16 % H_2O/Ar	573 K (30 min)
Desorption isothermal N_2 purge Temp. swing	Flow of the ads step Flow of the ads step	100 % N_2 100 % N_2	573 K (30 min) 573–673 K (10 K min^{-1}, 30 min)

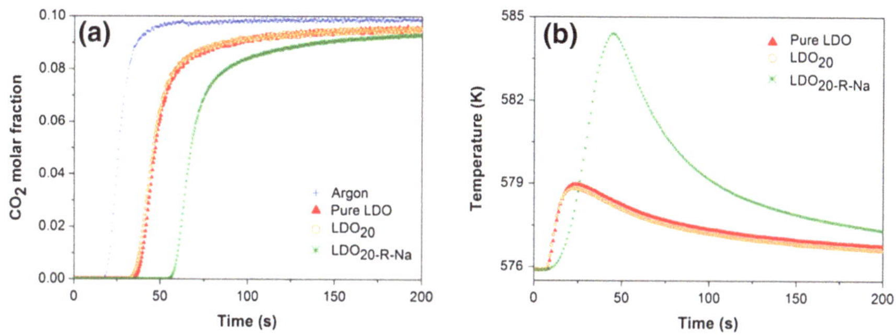

Fig. 7.1 **a** Breakthrough curves and **b** temperature profiles for pure LDO, LDO_{20} and $LDO_{20\text{-R-Na}}$ under dry conditions: 573 K, 1 bar, 10 % CO_2/Ar, 46 ml min^{-1} STP, $m_a = 0.3$ g

7.2.1 Dry Breakthrough Curve Tests

Figure 7.1a shows typical first-contact breakthrough curves[1] under dry conditions for a pure LDO, a LDO/GO hybrid with 5 wt% GO (LDO_{20}) and a promoted LDO/GO hybrid containing 5 wt% GO and 1.9 wt% residual sodium ($LDO_{20\text{-R-Na}}$).[2] Although the adsorption step was carried out for 30 min, the increase in the CO_2 concentration in the gas effluent was minimal after 400 s under the operating conditions used. The breakthrough curves of all the adsorbents show a sharp increment before the stoichiometric time of the column indicating fast initial adsorption kinetics. A more gradual uptake was observed afterwards as the CO_2 breakthrough approaches equilibrium. Similar CO_2 adsorption profiles on LDOs have been observed by other authors who have suggested that the slow adsorption stage close to saturation is related to internal mass transfer limitations and the non-isothermicity of the bed [1–4]. The fast and slow uptake processes were also observed by CO_2-TGA analysis at 573 K (Sect. 5.4, Chap. 5).

The adsorption capacities of the materials determined by fixed-bed and TGA measurements showed good agreement (Fig. 7.2), which validates the methodology used. The standard error of the breakthrough data increased slightly with the partial pressure of CO_2 in the feed possibly due to the sharper concentration wavefronts caused by the increase in the CO_2 molar flow rates. The pure LDO and the LDO_{20} hybrid show similar breakthrough times corresponding to adsorption capacities of 0.23 and 0.24 mol CO_2 kg ads^{-1} respectively (at 573 K, pCO_2 = 100 mbar). On the other hand, the breakthrough time of the $LDO_{20\text{-R-Na}}$ hybrid was significantly longer giving an adsorption capacity of 0.5 mol CO_2 kg ads^{-1}. These results corroborate the findings of CO_2-TPD and TGA tests (Sects. 6.3 and 6.4) showing that

[1]First-contact refers to the first exposure to the adsorptive gas after activation.

[2]The procedure to synthesise the adsorbents and their physicochemical properties are given in Chaps. 5 and 6.

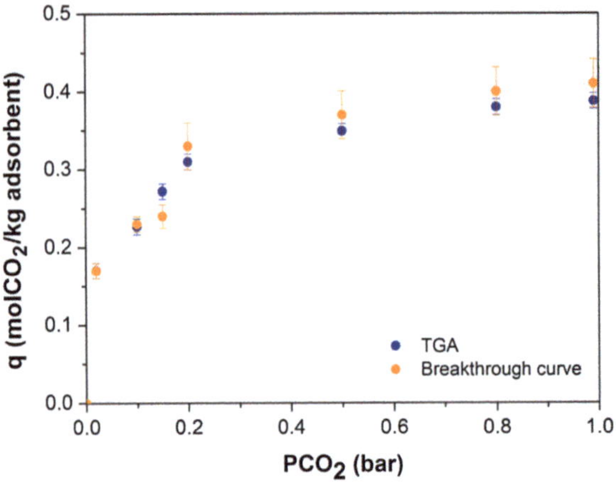

Fig. 7.2 First-contact adsorption isotherms at 573 K under dry conditions from TGA and breakthrough curve measurements for pure LDO; shown with standard errors

GO acts mainly as an efficient support whereas the presence of alkali metals alters considerably the basicity of the LDO adsorbents resulting in higher CO_2 uptakes.

The temperature measured in the middle of the packed bed reveals the existence of a heat wavefront caused by the exothermic adsorption of CO_2 on the materials, Fig. 7.1b. Under dry conditions the unpromoted samples, either pure or GO supported, show a temperature increment of ca. 3 K. In contrast, the maximum of the temperature peak for the sodium promoted hybrid corresponds to an increase of ca. 8 K. The higher release of heat during the adsorption of CO_2 on the $LDO_{20\text{-}R\text{-}Na}$ is related to its higher capacity (approximately double that of pure LDO) and to a slight increase in the average strength of the surface-adsorbate interactions (see the discussion on the CO_2-TPD experiments in Sect. 6.3).

The stability of the adsorption capacity was explored in both isothermal N_2 purge and temperature-swing fixed-bed experiments. The breakthrough curves of LDO_{20} in multiple adsorption-desorption cycles are presented in Fig. 7.3. Similar

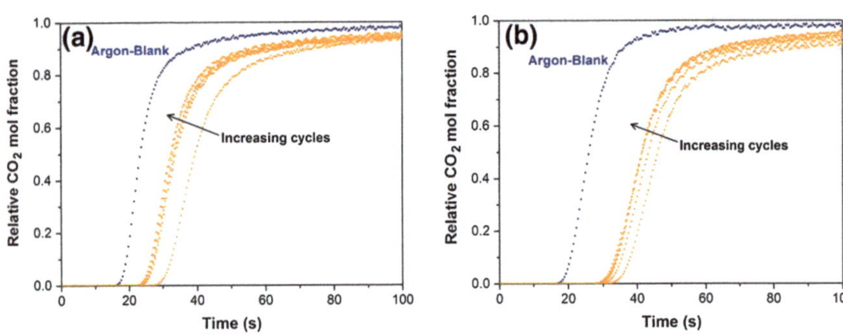

Fig. 7.3 **a** Isothermal N_2 purge (six cycles; adsorption 573 K; desorption 573 K) and **b** temperature-swing (six cycles; adsorption 573 K; desorption 673 K) for LDO_{20} under dry conditions: 1 bar, 10 % CO_2/Ar, 46 ml min^{-1} STP, m_a = 0.3 g

profiles were obtained for the pure LDO and $LDO_{20\text{-R-Na}}$ and are given in Appendix F.1. In isothermal N_2 purge and temperature-swing cycles, there is a significant loss of capacity during the first-contact-adsorption which is caused by the irreversible chemisorption of CO_2 over the strongest basic sites created during the activation (see discussion in Sect. 5.4, Chap. 5). In the isothermal N_2 purge experiment almost all the loss in capacity occurs in the first-contact. The capacity recovered in the first temperature-swing cycle is higher than in the first isothermal N_2 purge cycle due to the thermal activation of the desorption process. From the second temperature-swing cycle, there is a progressive small decrease in the adsorption capacity, which may be due to the gradual ripening of the LDO particles in a process that is likely to be temperature-driven. After the first cycle in isothermal N_2 purge experiments, the adsorption capacity is more stable and mainly weak-chemisorption sites are cyclically occupied and evacuated. Other authors have reached the same conclusion from the values of the heat of adsorption calculated between 673 and 793 K [1, 5, 6]. It is worth mentioning that the stability trends observed in thermogravimetric and fixed-bed TSA experiments were similar.

7.2.2 *Wet Breakthrough Curve Tests*

A breakthrough curve for the LDO_{20} hybrid (5 wt% GO) in the presence of steam (0.16 mol H_2O mol^{-1} total) is shown in Fig. 7.4a. A similar profile was observed for the wet pure LDO. As in the case of the dry experiments, the wet curves show fast initial kinetics followed by a slower CO_2 uptake process, approaching the inlet CO_2 concentration in less than 400 s. The wet adsorption capacities of the materials were found to be 10–40 % higher than those obtained under dry conditions over the range of CO_2 partial pressures tested. A thermogravimetric analysis of LDO_{20} hybrids recovered after wet adsorption gave the same carbon content as the fresh materials. This indicates that gasification of GO does not occur to any significant extent in the presence of water under the conditions tested.

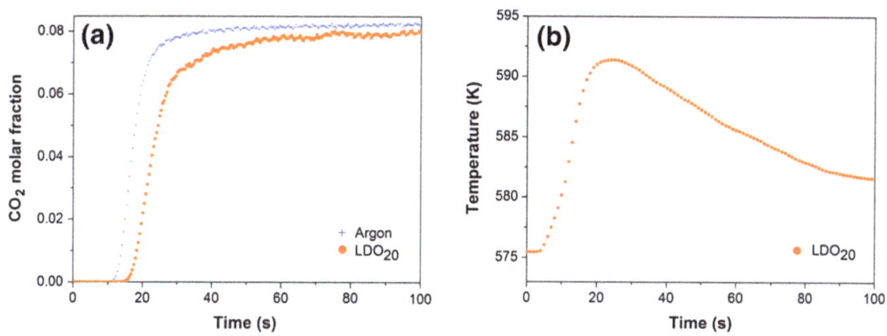

Fig. 7.4 **a** Breakthrough curve and **b** temperature profile for LDO_{20} under wet conditions: 573 K, 1 bar, 8.4 % CO_2/ 16 % H_2O/Ar, 165 ml min^{-1} STP, $m_a = 0.3$ g

The exposure to H_2O and CO_2 was observed to produce an increase in the temperature inside the column ($\Delta T = 21$ K) that is considerably higher than under dry conditions ($\Delta T = 4$ K), suggesting substantial adsorption of water by the materials, Fig. 7.4b. Regrettably, due to the high H_2O concentration used in the present experiments, it was necessary to place a water trap before the sampling point to obtain stable readings in the mass spectrometer. Consequently, it was not possible to follow the water breakthrough and to measure the total adsorptive uptake. In any case, the fact that water causes an increase in the CO_2 capacity indicates that the exothermic surface rehydration leads to the creation of new adsorption sites, possibly aluminium and magnesium hydroxides as proposed by Ram Reddy et al. [7]. Rehydration of LDOs has been reported to enhance the catalytic activity in several base-catalysed reactions for which OH^- acts as a Brønsted base site [8]. A similar effect is expected for the CO_2 adsorption capacity of LDOs as reported by other authors. Some examples are given in Table 7.2.

The first-contact adsorption isotherms under wet conditions for the pure LDO and the LDO_{20} hybrid were obtained from breakthrough analysis. The experimental data were fitted to the Langmuir, Freundlich (Fig. 7.5, Table 7.4) and Toth models (Fig. 7.6; Table 7.3) where carbon dioxide is regarded as the only adsorbed species.

Table 7.2 Adsorption capacities of various LDO based materials

Authors	Adsorbent	Support (wt%)	mol CO_2 kg^{-1} LDO	mol CO_2 kg^{-1} ads	Conditions
Present research	LDO	0	0.23	0.23	Dry, 573 K, pCO_2 = 100 mbar
Present research	LDO	0	0.28	0.28	Wet, 573 K, pCO_2 = 84 mbar
Present research	LDO_{20}	0	0.24	0.24	Dry, 573 K, pCO_2 = 100 mbar
Present research	LDO_{20}	0	0.29	0.29	Wet, 573 K, pCO_2 = 84 mbar
Alpay et al. [1]	LDO_{-K}	0	0.40	0.40	Dry, 673 K, pCO_2 = 100 mbar
Alpay et al. [9]	LDO_{-K}	0	0.44	0.44	Wet, 673 K, pCO_2 = 100 mbar
Hufton et al. [10]	LDO_{-K}	0	0.52	0.52	Dry, 673 K, pCO_2 = 700 mbar
Hufton et al. [10]	LDO_{-K}	0	0.85	0.85	Wet, 673 K, pCO_2 = 300 mbar
Costa et al. [7]	LDO	0	0.61	0.61	Dry, 473 K, pCO_2 = 1000 mbar
Costa et al. [7]	LDO	0	0.71	0.71	Wet, 473 K, pCO_2 = 880 mbar
Meis et al. [11]	LDO/CNF	CNF (90 wt%)	1.3	0.13	Wet, 523 K, pCO_2 = 50 mbar

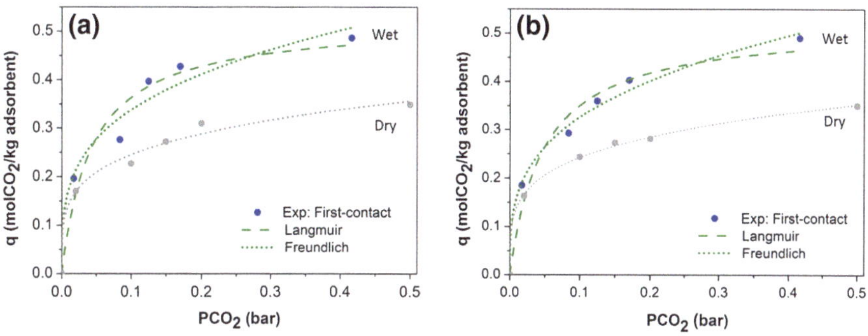

Fig. 7.5 First-contact adsorption isotherms at 573 K under wet conditions obtained from breakthrough analysis for **a** pure LDO and **b** LDO_{20}; data shown with standard errors. Dry data (Freundlich) from TGA

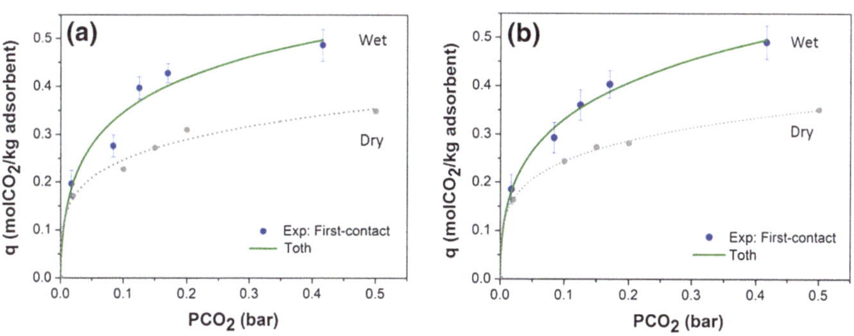

Fig. 7.6 First-contact adsorption isotherms at 573 K under wet conditions obtained from breakthrough analysis for **a** pure LDO and **b** LDO_{20}; data shown with standard errors. Dry data (Toth) from TGA

As shown in Table 7.4, the magnitude of the adjusted parameters is comparable to previously reported values. The Freundlich isotherm was found to describe slightly better the equilibrium data for both adsorbents, although a clear distinction is difficult due to the magnitude of the error bars. To clarify the type of adsorption, experiments need to be conducted in a wider range of CO_2 partial pressure using higher loadings of adsorbent in the column. In addition, it is important to

Table 7.3 Toth parameters for first-contact isotherms at 573 K under wet conditions

Adsorbent	Toth isotherm[a]			
	m (mol kg^{-1})	b (bar^{-1})	t (–)	R^2
Pure LDO	1.14 ± 2.02	199.4 ± 964.7	0.29 ± 0.45	0.95
LDO_{20}	2.68 ± 5.70	1003.8 ± 5084.8	0.18 ± 0.18	0.99

[a]The parameters are shown with the standard errors obtained from the fittings

Table 7.4 Parameters for adsorption isotherms from LDOs at 573 K reported under wet conditions

Authors	Adsorbent	Langmuir isotherm[a]			Freundlich isotherm[a]		
		m (mol kg^{-1})	b (bar^{-1})	R^2	k (mol kg^{-1})	n (–)	R^2
Present work (first-contact)	LDO	0.52 ± 0.05	23.1 ± 9.6	0.95	0.65 ± 0.06	3.50 ± 0.63	0.97
Present work (first-contact)	LDO_{20}	0.52 ± 0.05	20.9 ± 7.0	0.97	0.65 ± 0.03	3.31 ± 0.26	0.99
Ding and Alpay et al. [1]	$LDO_{\text{-K}}$	0.65	40.1[b]	–	–	–	–
Halabi et al. [4]	$LDO_{\text{-K}}$	–	–	–	1.11[c]	1.73[c]	–
Reijers et al. [12]	$LDO_{\text{-K}}$	–	–	–	0.70[c]	3.3[c]	–

[a]The parameters are shown with the standard errors obtained from the fittings
[b]The value of b at 573 K was estimated using the heat of adsorption reported in the reference
[c]Parameters reported at 673 K

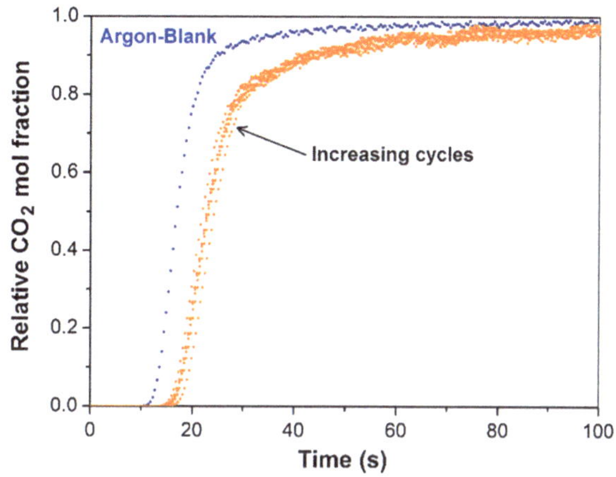

Fig. 7.7 Temperature-swing (six cycles; adsorption 573 K; desorption 673 K) for LDO_{20} under wet conditions: 1 bar, 8.4 % CO_2/16 % H_2O/Ar, 165 ml min^{-1} STP, $m_a = 0.3$ g

further investigate the role of water. Some authors [1, 12] have observed that the partial pressure of water has little effect on the CO_2 capacity of LDOs and therefore the non-competitive adsorption models here presented are considered to be a good approximation to the real adsorption isotherms. Recently, Boon et al. [13] investigated the adsorption of CO_2 and H_2O on a K-promoted LDO and found that a multicomponent adsorption isotherm describes adequately equilibrium data at 673 K and in a partial pressure range between 0.5 and 24 bar. The isotherm consists of two terms: a low partial pressure contribution in which H_2O and CO_2 adsorb on different surface sites, and a high pressure competitive physisorption due to the filling of nanopores.

The stability of LDOs under wet conditions was assessed by temperature-swing cycling. Representative breakthrough curves are shown in Fig. 7.7 and Appendix F.2. Compared to experiments under dry conditions, a significant decrease in the deactivation rate is observed in the presence of steam. Similar findings have been reported by Costa et al. [7] and Ding et al. [1]. It is possible that water mitigates the loss of capacity by maintaining the hydroxyl concentration of the surface and/or preventing the poisoning by carbonate formation or by coke deposition [1].

7.3 Mathematical Modelling

The first step towards the mathematical description of the adsorption process in the packed column is to model the non idealities of the flow caused by molecular and turbulent diffusion and by the viscous characteristics of the real fluid. To this end, the mixing that occurs in the system was assessed from step input experiments using argon as a tracer. A typical argon breakthrough curve and the plot of the residence-time distribution function, E(t), against time are presented in Fig. 7.8.

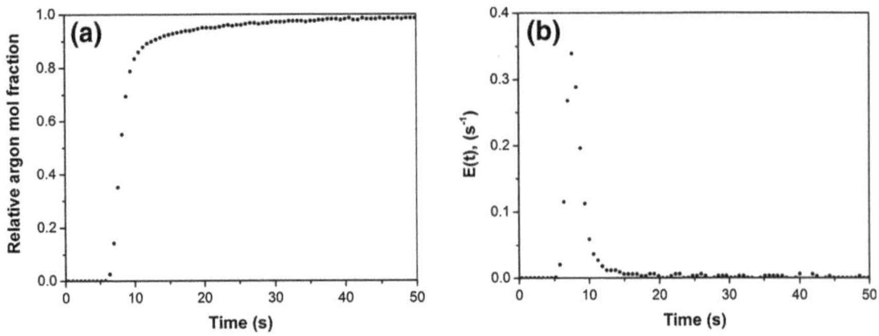

Fig. 7.8 **a** Argon breakthrough curve and **b** residence-time distribution (t_m = 8.1 s, σ^2 = 1.6 s^2, s^3 = 19)

The residence-time distribution (RTD) moments are given in the figure caption. The magnitude of the variance, $\sigma^2 > 0$, indicates that the tracer response is spread with respect to the mean. In fact, the positive value of the third moment of the RTD, s^3, implies that the distribution is slightly skewed to the left due to back-mixing.

To account for the deviation from ideal plug flow, two different approaches can be followed: (a) an apparent diffusivity can be defined so that a diffusion-type equation can be superposed to the convective transport or (b) the ideal plug flow can be assumed to be in series with a number of mixing-cells, i.e. perfectly stirred tanks. It has been reported that for small deviations from ideal plug flow both models give satisfactory results and the approach chosen is largely a matter of personal preference [14]. However, if a diffusion-type equation is to be applied to the profile presented in Fig. 7.8, it would be necessary to consider the dispersion characteristics of the different sections through which the gases flow, i.e. tubing lines, SiC pre-packing, quartz wool dividers, steel rod, fluid components, etc. Conversely, the tanks-in-series model is relatively easy to implement mathematically and provides a good approximation to the residence time distribution. Therefore, the transient mass balance equations for the tracer were written as:

$$\text{Plug Flow Model (PF)}: \quad \frac{\partial c_{Ar}}{\partial t} = -Q_f \frac{\partial c_{Ar}}{\partial V_{PF}} \tag{7.1}$$

$$\text{Continuous-Stirred Tank (CST)}: V_{CST} \frac{\partial c_{Ar}}{\partial t} = Q_f \left(c^o_{Ar} - c_{Ar} \right) \tag{7.2}$$

where t is the time counted from the instant when the step input is applied, c_{Ar} is the time-dependent concentration of the argon tracer, c^o_{Ar} is the concentration of argon in the CST feed, V_{PF} and V_{CST} are the volume of the corresponding ideal vessels, and Q_f is the volumetric flow at the conditions of the breakthrough experiment.

The volumes V_{PF} and V_{CST} are unknown and need to be estimated but the total dead-volume ($\varepsilon_T V_b$) can be obtained from the argon breakthrough curve as

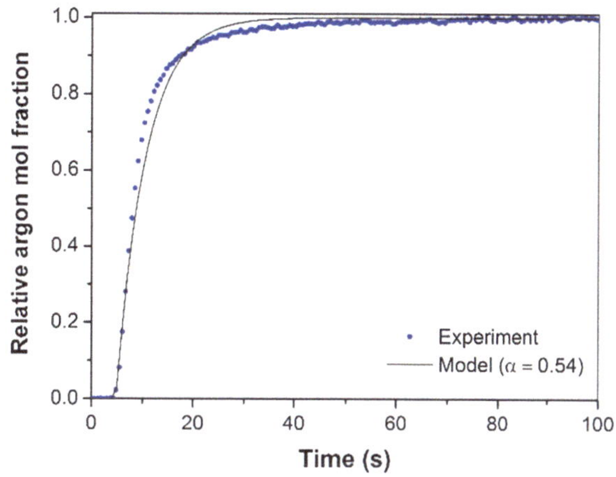

Fig. 7.9 Argon breakthrough curve: comparison between experimental data and mathematical model (CST + PF series, α = 0.54)

explained in Sect. 3.2.2.3. This dead-volume can be divided into the two components (plug flow and back-mixing) so that

$$V_{CST} = \alpha(\varepsilon_T V_b) \tag{7.3}$$

$$V_{PF} = (1 - \alpha)(\varepsilon_T V_b) \tag{7.4}$$

The value of the parameter α for each experiment was estimated from an argon blank using gPROMS. A curve fitted to the tracer concentration in the effluent is presented in Fig. 7.9 as an example. It is worth noting that the calculated dead-volumes do not correspond necessarily to the physical volume in the column. Most of the dead volume is associated with the fluid components to the detector and with the sampling system.

7.3.1 Adsorption Column Model

Separations by adsorption are usually carried out in fixed-bed columns and it is of great interest to have models that predict the adsorptive recovery for a given set of conditions, i.e. flow velocity, concentration, temperature, time and location in the bed. Such mathematical expressions need to account for the heat and mass balances (including the resistance to transport), the fluid dynamics and the thermodynamics of the adsorption. In this research, the equations used to describe the CO_2 breakthrough curves obtained experimentally were based on the following assumptions:

- A one-dimensional axially dispersed plug flow model was used. The velocity in each differential (axial) section is constant at its mean value. Likewise, the radial gradients of concentrations and temperatures are considered to be negligible. Usually the radial dispersion is regarded as unimportant when the tube diameter is far greater than the particle diameter [15]. As a rule of thumb a d_t to d_p ratio above 12 is recommended; [16] for the present experiments it was 17.

- The velocity of the fluid does not vary over time, i.e. steady-state approximation for the momentum balance.
- The magnitude of the Biot numbers presented in Sect. 3.2.2.4 indicate that the interparticle resistance to transport is minimal compared to intraparticle limitations. Consequently the effect of external mass transport is not included in the mathematical description.
- The dispersion of the fluid through the adsorbent bed is modelled by a diffusion-type equation while the back-mixing in the rest of the lines (i.e. the main contribution to the observed dispersion) is described by ideal plug flow and stirred vessels in series.
- The adsorption kinetics are described by the linear driving force (LDF) approximation.
- Due to the high dt/dp ratio (around 17) of the column, the temperature variations in the radial direction can be neglected. In general it has been observed that the radial conductivity (λ) decreases strongly in the vicinity of the wall. This is caused by variations in the packing density and the flow velocity near the wall. As the particle size decreases with respect to the column diameter, the variations in the bed structure become less relevant and the one dimensional model results in a good prediction.
- Heat is constantly supplied to the packed bed through the tube wall. The adsorbent bed is assumed to consist of identical spherical particles of 464 μm which corresponds to the mean particle size determined in a Malvern particle analyser.
- All the species in the gas phase behave as ideal gases. Under dry conditions, carbon dioxide is the only adsorbed species.
- The Langmuir, Freundlich and Toth isotherms fit adequately to the experimental data of adsorption equilibrium.

The transient conservation equations for mass and energy and the steady-state momentum balance can be written as follows:

$$\varepsilon_T \frac{\partial c_i}{\partial t} + \frac{\partial (u c_i)}{\partial z} = \varepsilon_b \frac{\partial}{\partial z}\left(D_z \frac{\partial c_i}{\partial z}\right) - \rho_{b,ads} \frac{\partial q_{CO_2}}{\partial t} \tag{7.5}$$

$$\left(\varepsilon_T c C_{v,gas} + \rho_{b,ads} C_{p,ads}\right) \frac{\partial T}{\partial t} = \frac{\partial}{\partial z}\left(k_z \frac{\partial T}{\partial z}\right) - c C_{p,gas} u \frac{\partial T}{\partial z} - \rho_{b,ads} \Delta H_{ads,CO_2} \frac{\partial q_{CO_2}}{\partial t} + \frac{4U}{d_t}(T_w - T) \tag{7.6}$$

$$\frac{\partial P}{\partial z} = -K_D u - K_V u^2 \tag{7.7}$$

where the Ergun equation is used to describe the pressure drop due to the friction losses of the gas flowing through the packed bed. Experimentally, the step input consisted in an instantaneous change from nitrogen to a mixture of carbon dioxide and argon. Accordingly, the initial conditions (t = 0, $z \in$ (0, L)) considered are: $c_{N_2} = c_{N_2,feed}$, $c_{CO_2} = 0$, $c_{Ar} = 0$, $q_{CO_2} = 0$ and $T = T_{feed}$. For the boundary conditions, the following equations were used:

Inlet conditions (z = 0)

$$-\varepsilon_b D_z \frac{\partial c_i}{\partial z} = u\left(c_{i,feed} - c_i\right) \tag{7.8}$$

$$-k_z \frac{\partial T}{\partial z} = u c_{feed} C_{p,gas}\left(T_{feed} - T\right) \tag{7.9}$$

$$P = P_{feed} \tag{7.10}$$

Outlet conditions (z = L)

$$\frac{\partial c_i}{\partial z} = 0 \tag{7.11}$$

$$\frac{\partial T}{\partial z} = 0 \tag{7.12}$$

$$uP = u_{feed} P_{feed} \tag{7.13}$$

The linear driving force model, LDF, is given by

$$\frac{\partial q_i}{\partial t} = k_{LDF}\left(q_i^* - q_i\right) \tag{7.14}$$

where k_{LDF} is the effective intraparticle mass transfer coefficient.

The adsorption equilibrium was described using one of the following three equations:

$$\text{Langmuir isotherm } q^*_{CO_2} = \frac{m b p_{CO_2}}{1 + b p_{CO_2}} \tag{7.15}$$

$$\text{Freundlich isotherm } q^*_{CO_2} = k p_{CO_2}^{1/n} \tag{7.16}$$

$$\text{Toth isotherm } q^*_{CO_2} = \frac{m b p_{CO_2}}{\left(1 + (b p_{CO_2})^t\right)^{1/t}} \tag{7.17}$$

The set of equations presented were solved using gPROMS (Process System Enterprises Ltd.). A backward finite difference method with first order collocation

Table 7.5 Parameters and operating conditions used in the simulations

Parameter	Value	Units	Comments
ε_b	0.40	(-)	Dixon et al. [17]
d_t	7.7	mm	Measured
L_{column}	364	mm	Measured
L_{bed}	20	mm	Measured
D_m	4.9E-05	$m^2\ s^{-1}$	Fuller et al. [18] and Blanc's et al. [19]
D_z	4.6E-05	$m^2\ s^{-1}$	Edwards et al. [20]
μ_{gas}	4.2E-05	$kg\ m^{-1}\ s^{-1}$	Chung et al. [19]
$C_{v,gas}$	12.9	$J\ K^{-1}\ mol^{-1}$	Value at 573 K
$C_{p,gas}$	21.3	$J\ K^{-1}\ mol^{-1}$	Value at 573 K
$C_{p,ads}$	850	$J\ K^{-1}\ kg^{-1}$	Ding et al. [1]
$\Delta H_{ads,CO2}$	−41,800	$J\ mol^{-1}$	TGA (average value)
k_g	0.03	$J\ m^{-1}\ s^{-1}\ K^{-1}$	Reijers et al. [12]
k_p	1.0	$J\ m^{-1}\ s^{-1}\ K^{-1}$	Reijers et al. [12]
k_z	0.13	$J\ m^{-1}\ s^{-1}\ K^{-1}$	Reijers et al. [12]
U	92	$J\ m^{-2}\ s^{-1}\ K^{-1}$	Reijers et al. [12]
K_D	157,704	$kg\ m^{-3}\ s^{-1}$	Ergun equation
K_V	29,820	$kg\ m^{-4}$	Ergun equation
T_w	573	K	
T_{feed}	573	K	
P_{feed}	1.0	bar	
u_{feed}	0.094	$m\ s^{-1}$	
y_i	0.02 CO_2, 0.98 Ar		

on 100 elements was found to give a converged solution with errors that did not exceed 1 %. A summary of the parameters used in the simulations is given in Tables 7.5 and 7.6.

7.3.2 *Verification of the Adsorption Model*

To test whether that the proposed mass balance equation is able to reproduce a typical CO_2 breakthrough profile and to check that no mistakes were made during the coding of the program, a simplified adsorption bed model was solved in gPROMS and compared to the analytical solution. For this assessment, the model presented in the previous section was further restricted as follows: (1) the adsorption equilibrium is described by Henry's law, $q^* = Hp_{CO_2}$, (2) the linear driving force coefficient has a constant value of 0.53 s^{-1}, (3) no axial dispersion occurs, and (4) the temperature, pressure and superficial velocity are constant.

The analytical solution for the simplified system is given by [12, 21]

Table 7.6 Parameters of pure LDO and LDO_{20} used in the simulations

Parameter	Value		Units	Comments
	LDO	LDO_{20}		
ε_p	0.48	0.47	(–)	Calculated from exp. data
ε_t	0.69	0.68	(–)	Calculated from exp. data
d_p	464	459	μm	Malvern particle size analyser
d_{pore}	10	13	nm	N_2 physisorption
$\rho_{b,ads}$	653	716	kg m^3	Measured
ρ_p	1222	1165	kg m^3	Calculated from exp. data
$\rho_{skeletal}$	2350	2190	kg m^3	Helium pycnometry
v_{pore}	3.9E-04	4.0E-04	m^3 kg^{-1}	N_2 physisorption
m	0.39	0.37	mol kg^{-1}	Langmuir isotherm
b	21.7	23.5	bar^{-1}	Langmuir isotherm
k	0.40	0.39	mol kg^{-1}	Freundlich isotherm
n	4.78	4.86	(–)	Freundlich isotherm
m	1.14	0.88	mol kg^{-1}	Toth isotherm
b	10,322	3394	bar^{-1}	Toth isotherm
t	0.17	0.21	(–)	Toth isotherm

$$\frac{y_{CO_2}(z)}{y_{CO_2,feed}} = \int_0^{\tau} \exp[-(\varphi+u)]I_0\left(2\sqrt{\varphi u}\right)du + \exp[-(\varphi+\tau)]I_0\left(2\sqrt{\varphi\tau}\right) \quad (7.18)$$

where I_0 is a zeroth-order modified Bessel function of the first kind and the variables φ and τ are defined as:

$$\varphi = \frac{k_{LDF}\rho_{b,ads}RTHz}{\varepsilon v} \quad (7.19)$$

$$\tau = k_{LDF}\left(t - \frac{z}{v}\right) \quad (7.20)$$

The analytical solution was evaluated using Matlab and was found to give the same results that the simulation in gPROMS, Fig. 7.10. The operating conditions and the parameters used in the calculations are given in Table 7.7.

Since the simplified mass balance is able to reproduce adequately the response of the system to a step input, the more complete model was solved and a preliminary sensitivity analysis of some parameters was conducted. In particular, the relevance of the effective diffusivity of the packed bed and the pressure drop equation were investigated. First, the axial effective diffusivity was obtained using the following correlation [20]:

$$D_z = 0.73D_m + \left(0.5d_p u\right)\left(\varepsilon_b + \frac{9.7\varepsilon_b^2 D_m}{d_p u}\right)^{-1} \quad (7.21)$$

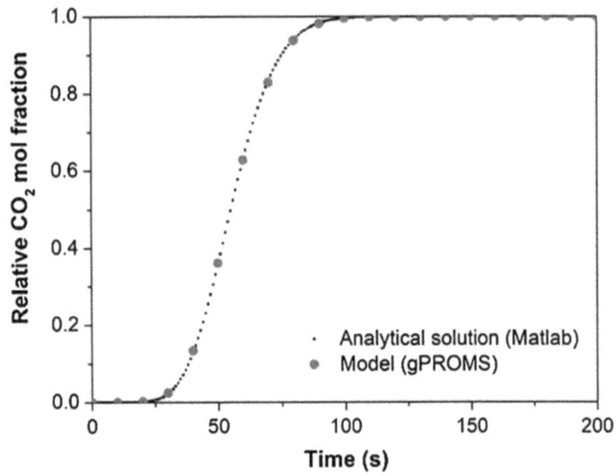

Fig. 7.10 Breakthrough curves comparison between analytical solution and a simplified model

The adsorption column was simulated with and without axial dispersion using the Eqs. (7.5–7.14), the Toth isotherm and a k_{LDF} of 0.08 s^{-1}. A comparison between both cases gave a maximum deviation in the CO_2 concentrations in the effluent of 2 %. Furthermore, a change in the effective diffusivity by ±20 % results in a maximum deviation with respect to the value from the correlation of 0.02 %, revealing that the system is only slightly sensitive to D_z.

It is worth noting that, under the conditions tested, the influence of the pressure drop equation on the CO_2 breakthrough profile was found to be negligible. In fact, for the maximum flow rate used the pressure drop through the adsorbent bed was calculated to be only 3 mbar. Consequently, it was possible to substitute the differential equation that describes the relation between the velocity and the pressure drop by an algebraic expression (7.22) that accounts only for the change in the linear velocity due to CO_2 adsorption.

$$u_{feed}\left(c_{Ar,feed} + c_{N_2,feed}\right) = u\left(c_{Ar} + c_{N_2}\right) \tag{7.22}$$

Table 7.7 Parameters and conditions used in the simulations

Parameters/conditions	Value	Units
k_{LDF}	0.53	s^{-1}
v	0.24	$m\ s^{-1}$
ε	0.40	–
ρ_{ads}	653	$kg\ m^{-3}$
H	8.5E−05	$mol\ kg^{-1}\ Pa^{-1}$
z	0.02	m
P	1	bar
y_i	0.02 CO_2, 0.98 Ar	–
T	573	K

7.3.3 *Adsorption Kinetics*

In this section, the parametric estimation of the linear driving force mass transfer coefficient is presented for adsorption experiments carried out using the pure LDO and the hybrid LDO_{20}. The data corresponds to first-contact adsorption for which isotherms were already available. Only dry experiments are included in the analysis since no information regarding the breakthrough of water was available due to the presence of a water trap. As a consequence the effect of competing species on the adsorption kinetics could not be assessed.

7.3.3.1 CO_2 Adsorption Kinetics Over Pure LDO

The adsorption experiments were conducted at an absolute pressure of 1 bar and 573 K using a premixed gas consisting of 2 % CO_2 balance in argon. Such low adsorptive concentrations were found to produce well defined breakthrough profiles and small temperature gradients that were convenient for the parameter estimation. The linear driving force mass transfer coefficients, k_{LDF}, fitted using the Langmuir, Freundlich and Toth isotherms are presented in Fig. 7.11. Also shown are simulated curves based on k_{LDF} values that are larger and smaller than the optimal estimated value.[3] For the three adsorption equilibrium models used, the steepness of the breakthrough curve is observed to increase with the value of k_{LDF}, as expected for less limited mass transport processes.

Since the data in Fig. 7.11 corresponds to a first-contact adsorption experiment, the evolution of the CO_2 concentration in the effluent is better predicted with the Freundlich or Toth equations than with the Langmuir expression. The optimal k_{LDF} for the Freundlich and Toth fittings is in the same order of magnitude that the estimated by Ding et al. [1] from experimental data and the calculated by Reijers et al. [12] based on a theoretical approach. It is important to stress that even for the best fitting, i.e. Freundlich isotherm and $k_{LDF} = 0.06\ s^{-1}$, the simulated approach of the CO_2 breakthrough to the inlet concentration is faster than what is experimentally observed. Moreover, the model also fails to match the temperature profile of the bed (Appendix F.3). Similar deviations between the model and the experiments were reported by Sircar et al. [5] and were attributed to non-isothermic and non-adiabatic behaviours that result in a slow approach to the steady-state temperature. As the temperature of the bed decreases, a slower CO_2 uptake occurs causing the tailing observed in the experimental breakthrough curve in Fig. 7.11. Alternatively, transport limitations in the column may account for part of the differences observed with respect to the experimental data. Rodrigues et al. [22], for instance, combined

[3]The optimal estimated value of k_{LDF} was found with the gPROMS in-built parameter estimation facility based on a Maximum Likelihood formulation. The 95 % t-value was $\gg$ 95 % reference t-value in all cases. In addition, the confidence intervals were $\ll$ respective estimated parameter value.

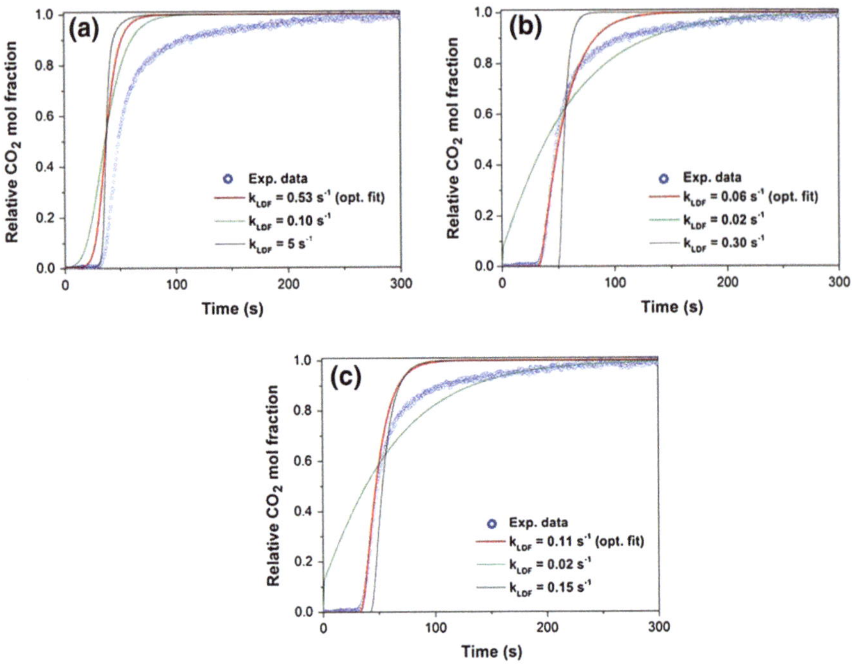

Fig. 7.11 CO_2 breakthrough curves for pure LDO (CO_2 outlet/CO_2 feed); 573 K, 2 % CO_2/Ar. Comparison between experiments and k_{LDF} model using **a** Langmuir isotherm, **b** Freundlich isotherm and **c** Toth isotherm

the heat balance equation inside the pellet with that of the surrounding fluid to include resistances to the dissipation of the heat generated by adsorption. Conversely, Ding et al. [1] found that the effect of heat transfer is minor compared to the influence of internal mass transport limitations. The authors reported an improvement in the fit to the experimental data by manipulation of the Knudsen diffusivity term in the theoretical expression for k_{LDF}.

A simultaneous optimisation of k_{LDF} and the Toth b parameter was conducted and the simulated breakthrough profile was compared to that obtained using the value fitted to the experimental isotherm ($b = 0.10\ \text{Pa}^{-1}$), Fig. 7.12. The fit of the model to the first-contact adsorption data was found to be sensitive to the value of b. It is likely that the difference between the experiment and the simulations is partly caused by the difficulty to fit the top region of the breakthrough profile and/or by the uncertainty in the parametric fitting of the isotherm as found by others [1].

7.3.3.2 CO_2 Adsorption Kinetics Over the LDO_{20} Hybrid

The experiments were conducted in the adsorption column under the same conditions of the tests carried out with pure LDO. The values of k_{LDF} estimated using the Langmuir, Freundlich and Toth isotherms are shown in Fig. 7.13. For the Langmuir

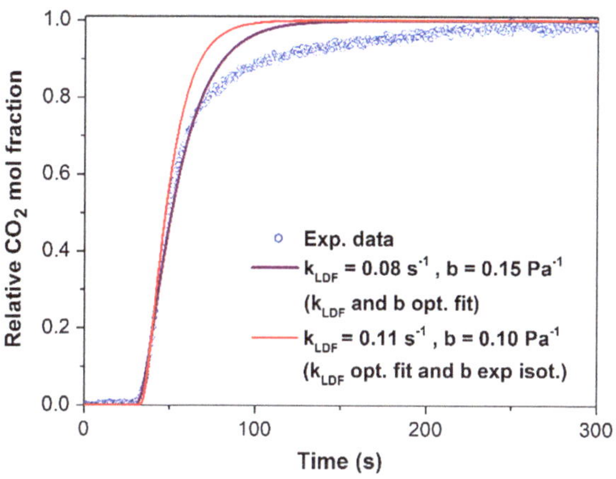

Fig. 7.12 CO_2 breakthrough curves for pure LDO (CO_2 outlet/CO_2 feed); 573 K, 2 % CO_2 (Ar). Sensitivity to the *b* parameter of the Toth equation

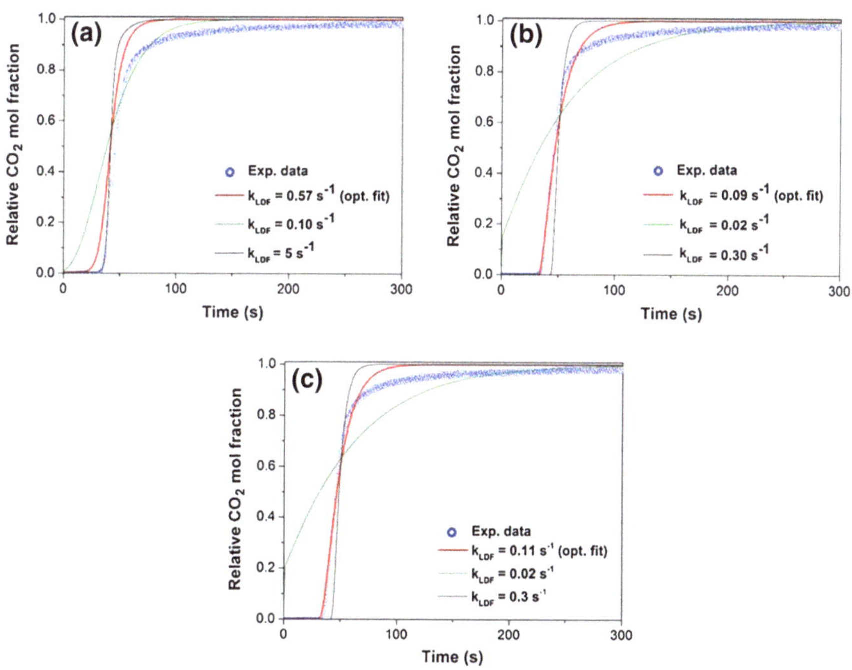

Fig. 7.13 CO_2 breakthrough curves for LDO_{20} (CO_2 outlet/CO_2 feed); 573 K, 2 % CO_2 (Ar). Comparison between experiments and k_{LDF} model using **a** Langmuir isotherm, **b** Freundlich isotherm and **c** Toth isotherm (b = 0.023 Pa^{-1})

and Freundlich equations the parameters fitted in Table 7.6 were used while a simultaneous optimisation of k_{LDF} and b was done in the case of the Toth expression.

Freundlich and Toth were found to give a better fitting to the experimental data than the Langmuir isotherm, although the difference is not as marked as in the case of the pure LDO. For the unsupported and GO supported adsorbents the estimated k_{LDF} values have the same order of magnitude, being slightly larger for the hybrids. However, the differences are relatively small considering the deviations between the experiments and the model and therefore it is impossible to reach any conclusion based on these observations. In any case, it seems that the presence of 5 wt% GO has little effect on the CO_2 adsorption kinetics. This result is in agreement with the similarity in the textural (N_2 physisorption and XRD) and chemical (CO_2-TPD) properties of pure LDO and the hybrids with low GO loadings.

7.3.3.3 Wet CO_2 Adsorption Kinetics Over the LDO_{20} Hybrid

The mass transfer coefficient for the linear driving force approximation was estimated from a wet CO_2 breakthrough curve experiment conducted over a sample of LDO_{20}. The adsorption equilibrium was described by the Toth isotherm based on the non-competitive adsorption of CO_2 and H_2O. As mentioned above, many authors have reported that for LDOs the CO_2 uptake does not depend strongly on the H_2O partial pressure and therefore expressions that do not include water in the adsorption terms are common [1, 23]. Even the isotherm recently published by Boon et al. [13], which includes the adsorption of both species, reduces to a non-competitive adsorption expression under the conditions of the present experiment.

The value of k_{LDF} fitted to the wet CO_2 breakthrough curve is close to the optimal values obtained from the dry experiments using LDO and LDO_{20} and is in agreement with the results of Ding and Alpay [1]. It seems that the structural

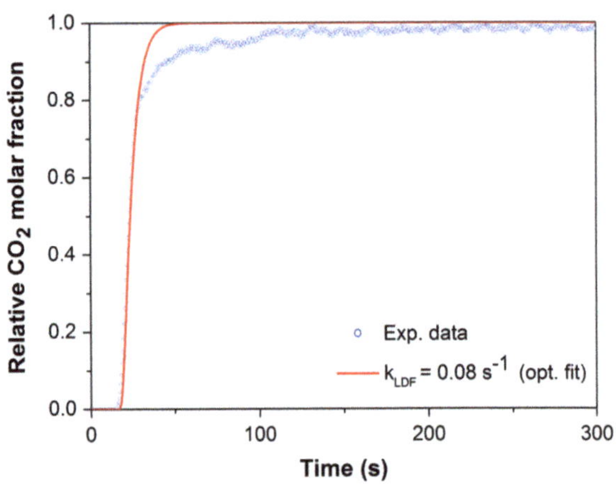

Fig. 7.14 Wet CO_2 breakthrough curve for LDO_{20} (CO_2 outlet/CO_2 feed); 573 K, 1.7 % CO_2/16 % H_2O/Ar. Comparison between experiment and k_{LDF} model using Toth isotherm

reconstruction that occurs upon rehydration of LDO does not impose severe restrictions to the diffusion of the adsorptive gas and therefore does not alter noticeably the kinetics of adsorption. As in the case of the dry tests, the model predicts a faster CO_2 breakthrough close to saturation compared to what is observed experimentally, Fig. 7.14.

7.4 Concluding Remarks

Unsupported and GO-supported LDO materials were studied in a fixed-bed column to obtain information on the equilibrium and the dynamics of CO_2 adsorption. The capacities determined from the fixed-bed experiments were found to be in close agreement with those measured by TGA, which validates the methodology used.

The breakthrough curves of all the materials tested exhibited a fast adsorption component in the beginning of the exposure to CO_2 followed by slightly slower uptakes as the effluent approaches the inlet concentration. Under dry conditions, the time-delay in the breakthrough of CO_2 due to adsorption is similar for the pure LDO and LDO_{20} but increases with the addition of sodium, as expected from the promoting effect of alkali ions on the equilibrium capacities. The increase in the bed temperature associated to the adsorption process is more marked for the Na promoted hybrid due to a combination of higher adsorption capacities and stronger bonding to the surface. The exposure of the adsorbents to a wet gas was found to be beneficial in terms of the measured CO_2 capacity, indicating that the adsorption of water results in the creation of new active sites. Additionally, the temperature increase measured in the bed (five times higher than the observed under dry conditions) suggests the occurrence of a highly exothermic process upon water adsorption, probably related to the partial reconstruction of the layered structure of Mg–Al hydroxides.

Multicycle temperature-swing and isothermal N_2 purge experiments indicated that the loss of CO_2 capacity after the first cycle is caused by irreversible adsorption over very strong basic sites that are created during the thermal activation. The adsorption capacity of the materials used in temperature-swing cycles continues to decay gradually, possibly indicating some additional thermal degradation. Water was shown to have a positive effect on the stability of unsupported LDO and the GO hybrids, therefore GO gasification during operation is unlikely.

The first-contact dry and wet CO_2 breakthrough profiles were adequately predicted by a nonisothermal one-dimensional transient model in which the adsorption equilibrium was well-described by the Toth or Freundlich isotherms. The linear driving force model (LDF) provides a good description of the adsorption kinetics up to a relative CO_2 capacity of $\sim 80\ \%$ but fails to predict the gradual breakthrough of CO_2 for higher surface coverages. However, since an industrial adsorption unit will be regenerated well before any CO_2 slip is observed, the crucial part of the breakthrough curve to be described is the fast initial uptake and this is in good agreement with the LDF model. The fit of the top part of the transient profile may be improved if limitations to the transport of energy are included or if the uncertainty in the parameters of the isotherm equation is decreased or if the pore diffusivity of the system is considered more thoroughly [1].

The experiments presented and discussed in this chapter expose some of the complexities of CO_2 adsorption on LDOs in which the presence of promoters or pre-adsorbed layers of water may modify the nature of the active sites increasing their equilibrium capacity and stability. However, efforts are still required to understand in detail some aspects regarding the competitive adsorption of species and its impact on the performance of the materials. The use of larger adsorption beds is expected to improve the quality of the experimental data. It is challenging to procure gradientless conditions in the gas phase using packed bed columns and therefore the use of a CST system may be useful to decrease mass and energy transport limitations of the CO_2 sorption process especially in the presence of water which is highly exothermic. So far CST adsorption systems have been rarely used for this purpose although their mathematical description is relatively easy to handle. Additionally, it is necessary to conduct adsorption tests using particles of different sizes since valuable transport information regarding the internal transport limitations can be extracted in this way. Likewise, the study of the mechanical properties of the LDO-based adsorbents including the inclusion of binders (e.g. Al_2O_3) or the use of structured packing is relevant for their commercial application. Usually gas adsorption separation is carried out with adsorbents shaped as pellets or granules containing a binder. However, mass transfer limitations and the possible existence of high pressure drops may impose constrains to the optimal operation of conventional packed bed columns. Novel adsorbent structures have appeared as an alternate approach to reduce the mass transfer resistance by decreasing the diffusion length without creating resistances to the flow. This is the case of non-particulate structure monoliths and laminate adsorbents. Investigation of these approaches may be useful for GO supported LDOs [24].

References

1. Ding, Y., & Alpay, E. (2000). Equilibria and kinetics of CO_2 adsorption on hydrotalcite adsorbent. *Chemical Engineering Science, 55*(17), 3461–3474.
2. Ebner, A. D., Reynolds, S. P., & Ritter, J. A. (2006). Understanding the adsorption and desorption behavior of CO_2 on a K-promoted hydrotalcite-like compound (HTlc) through nonequilibrium dynamic isotherms. *Industrial and Engineering Chemistry Research, 45*(18), 6387–6392.
3. van Selow, E. R., Cobden, P. D., Verbraeken, P. A., Hufton, J. R., & van den Brink, R. W. (2009). Carbon capture by sorption-enhanced water—gas shift reaction process using hydrotalcite-based material. *Industrial and Engineering Chemistry Research, 48*(9), 4184–4193.
4. Halabi, M. H., de Croon, M. H. J. M., van der Schaaf, J., Cobden, P. D., & Schouten, J. C. (2012). High capacity potassium-promoted hydrotalcite for CO_2 capture in H_2 production. *International Journal of Hydrogen Energy, 37*(5), 4516–4525.

5. Lee, K. B., Beaver, M. G., Caram, H. S., & Sircar, S. (2007). Chemisorption of carbon dioxide on sodium oxide promoted alumina. *AIChE Journal, 53*(11), 2824–2831.
6. León, M., Díaz, E., Bennici, S., Vega, A., Ordóñez, S., & Auroux, A. (2010). Adsorption of CO_2 on hydrotalcite-derived mixed oxides: Sorption mechanisms and consequences for adsorption irreversibility. *Industrial and Engineering Chemistry Research, 49*(8), 3663–3671.
7. Reddy Ram, M. K., Xu, Z. P., & Diniz da Costa, J. C. (2008). Influence of water on high-temperature CO_2 capture using layered double hydroxide derivatives. *Industrial and Engineering Chemistry Research, 47*(8), 2630–2635.
8. Debecker, D. P., Gaigneaux, E. M., & Busca, G. (2009). Exploring, tuning, and exploiting the basicity of hydrotalcites for applications in heterogeneous catalysis. *Chemistry—A European Journal, 15*(16), 3920–3935.
9. Ding, Y., & Alpay, E. (2000). Adsorption-enhanced steam–methane reforming. *Chemical Engineering Science, 55*(18), 3929–3940.
10. Hufton, J. R., Mayorga, S., & Sircar, S. (1999). Sorption-enhanced reaction process for hydrogen production. *AIChE Journal, 45*(2), 248–256.
11. Meis, N. N. A. H., Bitter, J. H., & de Jong, K. P. (2009). Support and size effects of activated hydrotalcites for precombustion CO_2 capture. *Industrial and Engineering Chemistry Research, 49*(3), 1229–1235.
12. Reijers, H. T. J., Boon, J., Elzinga, G. D., Cobden, P. D., Haije, W. G., & van den Brink, R. W. (2009). Modeling study of the sorption-enhanced reaction process for CO_2 capture. I. Model development and validation. *Industrial and Engineering Chemistry Research, 48*(15), 6966–6974.
13. Boon, J., Cobden, P. D., van Dijk, H. A. J., Hoogland, C., van Selow, E. R., & van Sint Annaland, M. (2014). Isotherm model for high-temperature, high-pressure adsorption of and on K-promoted hydrotalcite. *Chemical Engineering Journal, 248*, 406–414.
14. Levenspiel, O., & Bischoff, K. B. (1964). Patterns of flow in chemical process vessels. In B. Thomas, J. W. H. Drew & V. Theodore (Eds.), *Advances in chemical engineering* (Vol. 4, pp. 95–198). New York: Academic Press.
15. Yang, R. T. (1997). *Gas separation by adsorption processes*. London, UK: Imperial College Press (Vol. 1).
16. Kapteijn, F., & Moulijn, J. A. (2008). Laboratory catalytic reactors: Aspects of catalyst testing. In *Handbook of heterogeneous catalysis*. Germany: Wiley-VCH Verlag GmbH & Co. KGaA.
17. Dixon, A. G. (1988). Correlations for wall and particle shape effects on fixed bed bulk voidage. *The Canadian Journal of Chemical Engineering, 66*(5), 705–708.
18. Fuller, E. N., Schettler, P. D., & Giddings, J. C. (1966). A new method for prediction of binary gas-phase diffusion coefficients. *Industrial and Engineering Chemistry, 58*(5), 18–27.
19. Poling, B. E., Prausnitz, J. M., & O'Connell, J. P. (2004). *The properties of Gases and Liquids*. New York, USA: McGraw-Hill.
20. Edwards, M. F., & Richardson, J. F. (1968). Gas dispersion in packed beds. *Chemical Engineering Science, 23*(2), 109–123.
21. Ruthven, D. M. (1984). *Principles of adsorption and adsorption processes*. NY, USA: Wiley.
22. Dantas, T. L. P., Rodrigues, A. E., & Moreira, R. F. P. M. (2012). Separation of carbon dioxide from flue gas using adsorption on porous solids. *Greenhouse Gases—Capturing, Utilization and Reduction*.
23. Reijers, H. T. J., Boon, J., Elzinga, G. D., Cobden, P. D., Haije, W. G., & Brink, R Wvd. (2009). Modeling study of the sorption-enhanced reaction process for CO_2 capture. II. Application to steam-methane reforming. *Industrial and Engineering Chemistry Research, 48*(15), 6975–6982.
24. Rezaei, F., & Webley, P. (2010). Structured adsorbents in gas separation processes. *Separation and Purification Technology, 70*(3), 243–256.

Chapter 8
Sorption-Enhanced Methanol-to-Shift for H_2 Production: Thermodynamics and Catalyst Selection

8.1 Introduction

The LDH-based adsorbents developed in this research can find application in relevant processes such as the sorption-enhanced water gas shift (SEWGS). This technology aims to improve the efficiency of the hydrogen production by steam reforming of methane and by coal or biomass gasification. In the sorption-assisted process the water gas shift takes place in the presence of a solid adsorbent, which selectively removes CO_2 from the gas phase as it is formed. Consequently the thermodynamic constraints on the WGS are lifted thereby increasing the production of H_2.

There are many potential advantages associated with the implementation of SEWGS technologies. Water gas shift is normally carried out in two consecutive stages: a high temperature step (at 573–773 K over a Fe/Cr catalyst) to benefit from high reaction rates, and a low temperature step (at 473–543 K, over a Cu/Zn/Cr catalyst) with favourable thermodynamics but poor kinetics. The possibility to overcome the equilibrium limitations by means of CO_2 adsorption allows WGS to take place at temperatures between 573 and 773 where the kinetics are adequate. For instance, while the maximum conversion of CO at 673 K is ~90 % for the conventional process, values above 98 % can be obtained by removing 75 % of the CO_2 formed which is not too demanding for a solid adsorbent, Fig. 8.1a. Additionally, the decrease in the amount of CO_2 (product) in the gas phase results in a lower demand of H_2O (reactant) for a given equilibrium conversion, Fig. 8.1b. Consequently the steam usage can be significantly reduced leading to lower operation costs.

Besides sorption-enhancement, other process intensification strategies for the water gas shift have been discussed in the academic and the patent literature. Since WGS is an exothermic reaction, it has been proposed that the energy requirements can be reduced if the transformation is combined with endothermic processes such as steam-reforming of oxygenates [1, 2]. Recently Haldor Topsøe

D. Iruretagoyena Ferrer, *Supported Layered Double Hydroxides as CO_2 Adsorbents for Sorption-enhanced H_2 Production*, Springer Theses,
DOI 10.1007/978-3-319-41276-4_8

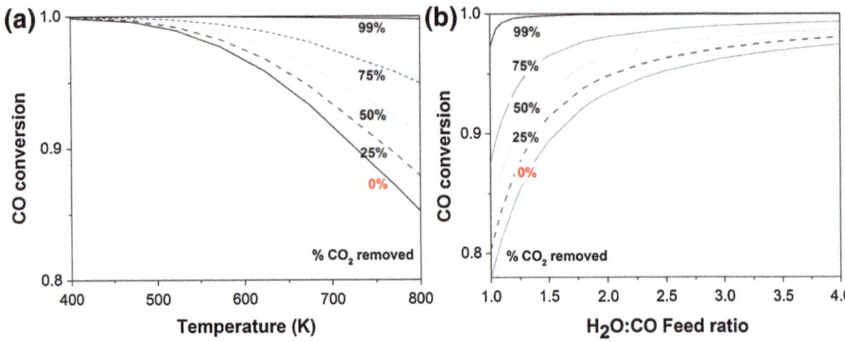

Fig. 8.1 SEWGS thermodynamic calculations **a** effect of temperature on CO conversion (H_2O:CO molar feed ratio = 2:1) and **b** effect of H_2O:CO feed ratio on CO conversion (at 673 K)

patented a process called methanol-to-shift (MTS) [3] which claims to increase the H_2 productivity and reduce the duty of the low temperature shift reactor by combining the WGS with the catalytic decomposition of methanol. The process is expected to be relevant in a context in which methanol is widely employed as energy carrier but it can be implemented using other oxygenates as feedstock. Furthermore, the methanol-to-shift process may be more competitive if it is combined with the in situ selective removal of the carbon dioxide formed.

Since there are no research works published that address the methanol-to-shift scheme, a deeper understanding is necessary to assess the viability of a sorption-enhanced process. Therefore, the present chapter aims to provide preliminary information on the thermodynamics of the system and to serve as guideline for future studies on water gas shift combined with the decomposition of methanol and other alcohols. In Sect. 8.2 the thermodynamics of methanol-to-shift are studied to evaluate the convenience of a sorption- assisted process. Subsequently, Sect. 8.3 presents the results of screening tests over catalyst systems that are promising candidates for the simultaneous decomposition of methanol and water gas shift. The catalysts that exhibited the best performance in the temperature range between 573 and 673 K, i.e. the operating window of LDOs, were tested for longer periods to assess their stability. Finally, concluding comments are given in Sect. 8.4.

8.2 Thermodynamic Analysis

A thermodynamic analysis was carried out to determine whether sorption assistance is useful for the methanol-to-shift process. The main criteria considered were the product selectivity, the hydrogen productivity and the energy efficiency of the process. A non-stoichiometric approach based on the principle of minimising the Gibbs free energy was used to find the equilibrium compositions of the system with

Table 8.1 Experimental data from examples 2, 10 and 17 [3]

Example 2		Example 10		Example 17	
T (K)	508	T (K)	586	T (K)	586
P (bar)	25	P (bar)	25	P (bar)	25
W_{cat} (g)	15.2	W_{cat} (g)	15.1	W_{cat} (g)	15.1
Feed (gas mole fraction)		Feed (gas mole fraction)		Feed (gas mole fraction)	
H_2	0.293	H_2	0.257	H_2	0.274
CO	0.097	CO	0.085	CO	0.091
CO_2	0.075	CO_2	0.068	CO_2	0.070
N_2	0.214	N_2	0.188	N_2	0.201
CH_3OH	0.039	CH_3OH	0.056	CH_3OH	0.000
H_2O	0.282	H_2O	0.346	H_2O	0.364
Total molar flow (mol h^{-1})	6.58	Total molar flow (mol h^{-1})	7.47	Total molar flow (mol h^{-1})	7.01
Exit (gas mole fraction)		Exit (gas mole fraction)		Exit (gas mole fraction)	
H_2	0.446	H_2	0.435	H_2	0.368
CO	0.008	CO	0.010	CO	0.008
CO_2	0.182	CO_2	0.176	CO_2	0.153
N_2	0.203	N_2	0.168	N_2	0.197
CH_3OH	0.008	CH_3OH	0.002	CH_3OH	0.000
H_2O	0.153	H_2O	0.209	H_2O	0.274
Total molar flow (mol h^{-1})	6.92	Total molar flow (mol h^{-1})	8.37	Total molar flow (mol h^{-1})	7.14
H_2 productivity (NL h^{-1})	1712	H_2 productivity (NL h^{-1})	2550	H_2 productivity (NL h^{-1})	1040

and without CO_2 in situ removal. A detailed description of the optimisation methodology is given in Appendix G.1.

The information provided in the patent filed by Haldor Topsøe [3] was used as a basis for the study as it is the most complete source among the scarce literature. The patent describes a process to produce a hydrogen-rich stream by co-feeding water gas and methanol in the presence of a catalyst comprising copper, zinc and aluminium oxides. The inlet temperature and pressure used were at least 553 K and 2.0 MPa respectively. Thermodynamic calculations of the methanol-to-shift reaction showed that the examples 2 and 10 from the patent, Table 8.1, were carried out close to chemical equilibrium as shown in Fig. 8.2. Therefore these runs were chosen as starting point to perform a sorption-enhancement analysis. The patent reports that the selectivity of the methanol conversion to carbon oxides and hydrogen was 100 % within the accuracy of the analytical techniques. Hence, for simulation purposes, it was assumed that the species coexisting in equilibrium were

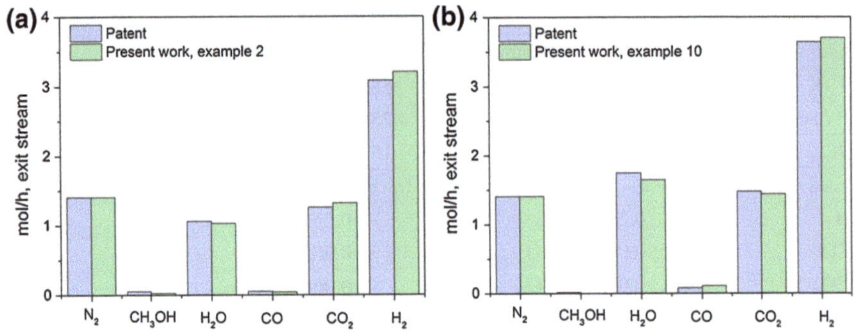

Fig. 8.2 Equilibrium molar flow: Comparison between equilibrium calculated in the present work and examples 2 (**a**) and 10 (**b**) from reference [3]

only CH_3OH, H_2O, CO, CO_2 and H_2. This implies that the following set of reactions take place[1]:

$$\text{Water gas shift:} \quad CO + H_2O \leftrightarrow CO_2 + H_2 \tag{8.1}$$

$$\text{Methanol decomposition:} \quad CH_3OH \leftrightarrow CO + 2H_2 \tag{8.2}$$

$$\text{Steam reforming of methanol:} \quad CH_3OH + H_2O \leftrightarrow CO_2 + 3H_2 \tag{8.3}$$

The effect of carbon dioxide removal on the equilibrium compositions under the operating conditions of example 10 is shown in Fig. 8.3. The concentration of hydrogen in the gas mixture increases with the fraction of carbon dioxide removed, whereas the rest of the species decrease. As expected, carbon dioxide adsorption shifts the equilibrium composition of the WGS reaction increasing the water conversion into hydrogen. In the ideal scenario in which the fraction of CO_2 adsorbed, f, is equal to 1, the H_2 mol fraction is nearly 1 (dry basis) while for the standard process, i.e. $f = 0$, the H_2 mol fraction is approximately 0.7.

Before moving further with the discussion of the sorption-enhanced methanol-to-shift (SEMTS) process, it is important to determine the effect of methanol addition to the system. For this purpose, the equilibrium compositions were calculated under the conditions of example 10 but without methanol in the feed (conditions of example 17), i.e. only the WGS reaction occurs. The results show that similarly to the sorption-enhanced methanol-to-shift, the sorption-assisted water gas shift leads to an effluent stream with high hydrogen purity (Fig. 8.4a). However, the hydrogen production by SEWGS is considerably lower than that of SEMTS. The simulation findings are consistent with the experimental results

[1]The reactions do not need to be specified for the calculation of the equilibrium composition using the Gibbs minimisation approach. However, they are helpful to understand the changes in the gas composition as the operating conditions vary.

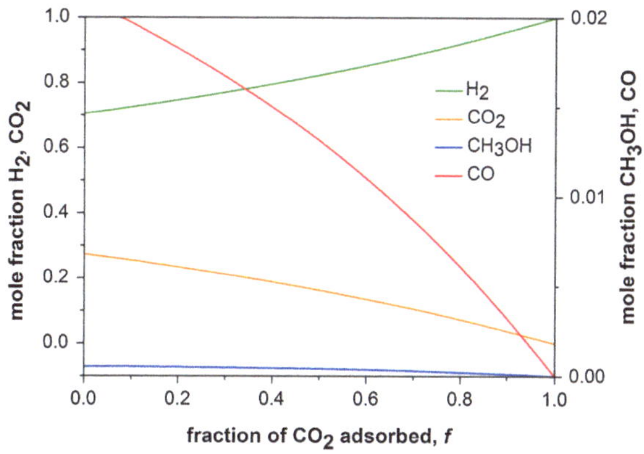

Fig. 8.3 Effect of CO_2 adsorption on the equilibrium compositions (dry basis), under the operating conditions of example 10

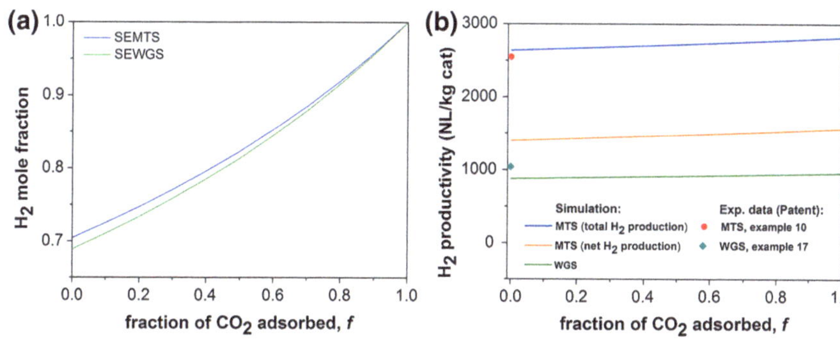

Fig. 8.4 **a** H_2 equilibrium mole fraction vs. *f*, SEMTS and SEWGS (operating conditions from examples 10 and 17 respectively) **b** H_2 equilibrium productivity versus *f*, SEMTS and SEWGS (operating conditions from examples 10 and 17 respectively)

included in the patent, Fig. 8.4b. As indicated in Table 8.1 and Fig. 8.4b, the H_2 production of example 17 (1040 NL H_2/h) is significantly improved in example 10 by the addition of methanol (2550 NL H_2/h).

The extra hydrogen produced by co-feeding methanol to a conventional WGS process comes partly from the hydrogen contained in the methanol molecules. Additionally, methanol decomposition produces carbon monoxide which reacts with the excess water to produce more hydrogen. This is clearly observed in Fig. 8.4b: the net H_2 production (discounting the H_2 inherent to the methanol molecules) is greater than the obtained by the WGS alone. The difference between the total and the net methanol-to-shift hydrogen production is twice the difference between the net MTS and the water gas shift alone. This is consistent with the stoichiometry of methanol decomposition which yields two H_2 molecules and one CO that is subsequently transformed into H_2 through WGS.

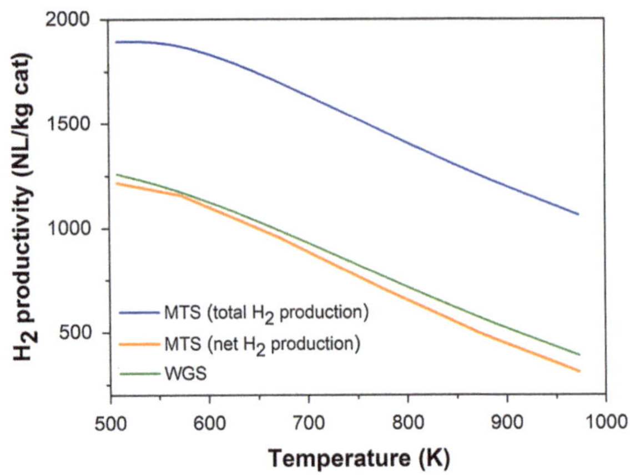

Fig. 8.5 Effect of the source of carbon on the hydrogen production. Standard methanol-to-shift and WGS reactions, operating conditions from example 2

Since the addition of methanol is an indirect way to feed carbon monoxide, the net hydrogen production obtained from the same amount of carbon, fed as CO (i.e. WGS) or as CH_3OH (i.e. methanol decomposition), were compared. As shown in Fig. 8.5, the net hydrogen production is slightly lower when H_2 is formed from the CO obtained by methanol decomposition than when the same amount of carbon monoxide is fed directly to the system.

To establish the temperature window in which sorption-assistance is useful for methanol- to-shift, the chemical equilibrium of the system was calculated from 373 to 1173 K using the feed compositions of example 2. The endothermic character of the methanol decomposition and the exothermicity of WGS are evidenced in Fig. 8.6. The exit molar flow of methanol decreases with temperature while carbon monoxide increases. Methanol decomposition is only slightly limited below 473 K, whereas WGS is significantly limited above 573 K. Therefore carbon dioxide adsorption becomes meaningful only above 573 K to overcome WGS equilibrium and a different catalyst from that reported in the MTS patent may be required for SEMTS. Above 573 K more hydrogen and less carbon oxides are produced in the sorption-enhanced process compared to the standard methanol-to-shift. As mentioned in the introduction to this chapter, high and low temperature shift reactors are industrially needed to overcome the thermodynamic and kinetic constraints of WGS, respectively. In this regard, sorption-enhanced methanol-to-shift can be carried out in a single stage since similar equilibrium conversions to those achieved by a conventional low temperature shift reactor (typically at 503 K) can be obtained at temperatures between 573 and 873 K.

The evaluation of the convenience of a new process should not be based only on the material balance but also on its energy efficiency. Consequently, the overall heat of reaction of the sorption-enhanced methanol-to-shift was determined using the feed composition of example 2 but at different methanol to CO ratios in the feed. Calculations were performed at 673 and 873 K, which are within the temperature range of interest. The standard molar enthalpies of all the species involved were retrieved from Smith et al. [4] and an enthalpy of adsorption of −10 kJ per mole of

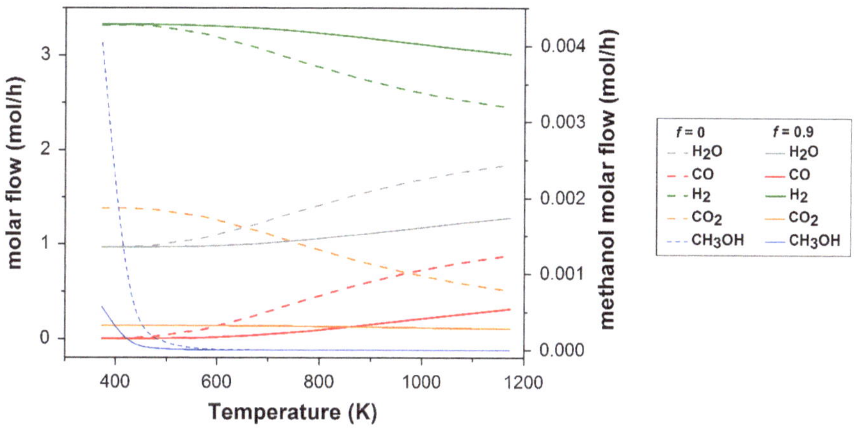

Fig. 8.6 Effect of temperature on the equilibrium molar flows of a methanol-to-shift system with and without CO_2 removal (1 bar, feed molar flows of example 2)

Fig. 8.7 Heat of reaction of SEMTS ($f = 0.99$) at different methanol to CO feed molar ratios at 673 and 873 K. The analysis was carried out at 1 bar using the feed molar flows of example 2 (except for methanol)

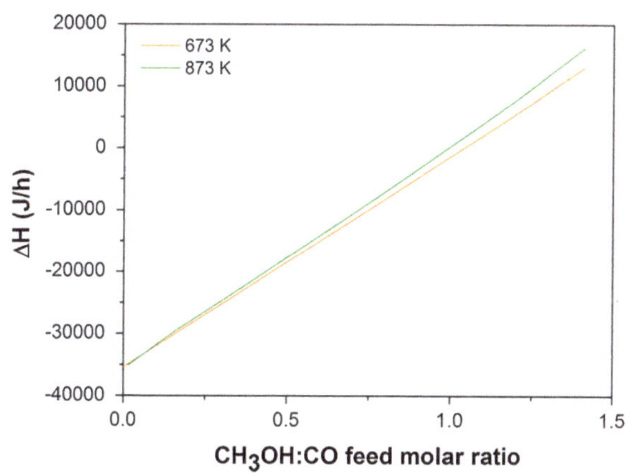

CO_2 was used [5]. As indicated in Fig. 8.7, there is a methanol to CO molar ratio in the feed that results in autothermal operation and is practically the same at 673 and 873 K: 1.03 and 0.98 respectively. The small difference is because the WGS conversion at 673 K is slightly higher compared to that at 873 K (even with total CO_2 removal), and therefore a larger amount of methanol needs to be fed. The overall heats of reaction of SEMTS and SEWGS were compared under the same operating conditions. As observed in Table 8.2 the former system is more energy efficient due the possibility to compensate the heat released by adsorption and the WGS reaction with the heat consumed by methanol decomposition.

Although methane was not detected in the experiments reported in the MTS patent in which a copper based catalyst was used, thermodynamically it is a very favoured species [6]. Therefore, the impact that the formation of methane would

Table 8.2 Comparison between SEMTS and SEWGS (f = 0.99), analysis at 1 bar using the molar flows of example 2 (except for methanol)

Process	T [K]	CH_3OH/CO feed molar ratio	H_2 exit molar flow (mol/h)	ΔHR (J/h)
SEMTS	673	1.03	4.53	−198
SEWGS	673	0.00	2.56	−35,391
SEMTS	873	0.98	4.41	−214
SEWGS	873	0.00	2.55	−35,070

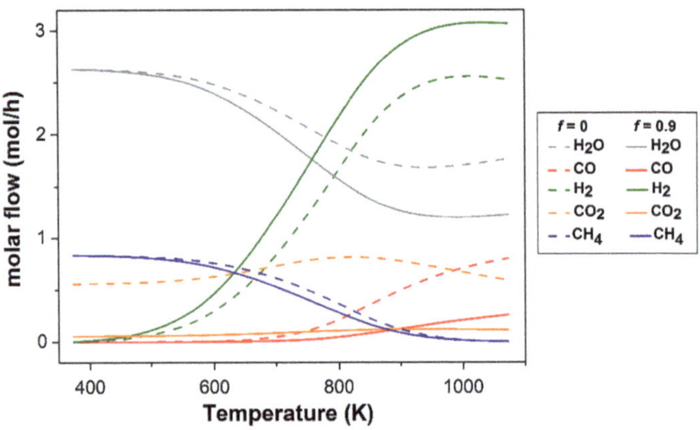

Fig. 8.8 Effect of temperature on the equilibrium molar flows of a methanol-to-shift system with and without CO_2 removal allowing for fast CH_4 formation (1 bar, feed molar flows of example 2). The corresponding molar flows of methanol are lower than 10^{-9} mol/h

have on the sorption-enhanced methanol-to-shift needs to be explored. The results in Fig. 8.8 indicate that methanation would reduce the quantity and quality of the hydrogen produced. However, similarly to a system without methane production, carbon dioxide adsorption increases the conversion of carbon monoxide and water to hydrogen above 573 K by shifting the WGS equilibrium composition.

Contrarily to the situation where methanation does not take place, at 673 K there is not a CH_3OH to CO feed ratio that enables autothermal operation when methane is produced at equilibrium, Fig. 8.9a. The carbon monoxide formed by methanol decomposition is mainly transformed into methane by the exothermic methanation reaction hence contributing to the overall heat released. On the other hand, at 873 K there is an optimum CH_3OH to CO feed ratio that minimises the overall heat of reaction of the SEMTS, Fig. 8.9b. In this case, methanol decomposition and methane steam reforming are thermodynamically favoured due to their endothermic character so part of the heat released by WGS and CO_2 adsorption can be compensated. Although at 873 K methane has a less negative effect than at 673 K, it is clear that a SEMTS process requires a selective catalyst that prevents methane formation.

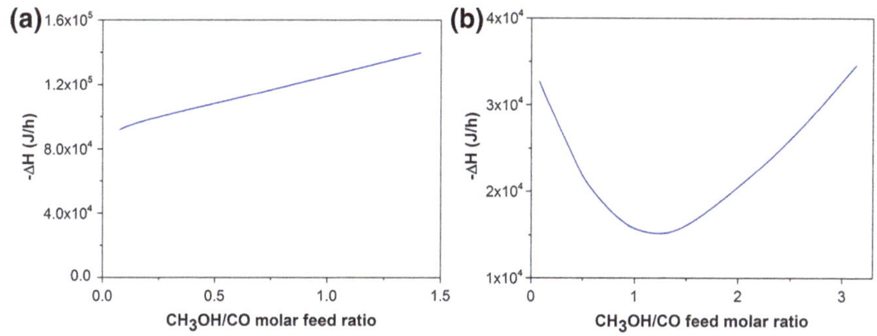

Fig. 8.9 Heat of reaction of SEMTS ($f = 0.99$) allowing for fast CH_4 formation at different methanol to CO feed molar ratios. The analysis was carried out at 1 bar using the feed molar flows of example 2 (except for methanol) **a** 673 K and **b** 873 K

8.3 Catalysts Screening

The suitability of different solid catalysts for the methanol-to-shift process was assessed by conducting the reverse water gas shift and methanol decomposition in temperature- programmed reaction experiments. An adequate catalyst should exhibit high activities and a minimum formation of hydrogenation by-products, e.g. methane, under the relevant operating temperatures (573–773 K). Commercial catalysts (Co/Mo, Fe/Cr, Cu/ZnO, Pt/Al_2O_3 and Pt/SiO_2) and in-house prepared catalysts (Pt/CeO_2, $Pt/CeO_2/SBA15$ and $Pt/CeO_2/ZrO_2$) were tested. The most promising catalysts identified in the fast screening were further studied in longer-term stability tests under water gas shift conditions.

8.3.1 Catalyst Synthesis and Characterisation

The in-house synthesised water gas shift/methanol decomposition catalysts were prepared by the incipient wetness method using commercial supports. Nanoceria powder, SBA-15 and ceria-zirconia were purchased from Sigma-Aldich, ACS Materials and Rhodia respectively. The supports were calcined flowing air at 773 K for 4 h before the impregnation. An aqueous solution of chloroplatinic acid, H_2PtCl_6, was used to impregnate the active metal. In the case of SBA-15, an intermediate impregnation step using an aqueous solution of $Ce(NO_3)_3{\cdot}6H_2O$ followed by calcination in air (773 K, 4 h) was conducted. After the supports were impregnated with platinum, they were dried at 373 K for 24 h and then calcined at 773 K under flowing N_2.

The catalysts selected for the study have been reported to be active for the water gas shift and/or methanol decomposition so their use in the methanol-to-shift reaction may be adequate [7–11]. The textural properties of the commercial and the

Table 8.3 Textural properties of the catalysts

Sample name	SBET (m^2/g)	Vpore (cm^3/g)	dpore (nm)
Co/Mo[a]	215	0.55	7.6
Fe/Cr[b]	176	0.21	4.8
Cu/ZnO[c]	84	0.26	9.9
Pt/Al_2O_3[d]	0.76	0.01	40.2
Pt/SiO_2[e]	253	1.05	12.0
Pt/CeO_2[f]	31	0.12	12.9
Pt/CeO_2/SBA15[g]	432	0.63	4.9
$Pt/CeO_2/ZrO_2$[h]	64	0.19	9.5

[a]Sour shift catalyst (Eurocat); typical composition: 3 wt% CoO, 11 wt% MoO_3, 72 wt% Al_2O_3, 14 wt% K_2CO_3
[b]High temp. shift catalyst (Alfa Aesar): typical composition: 80–90 wt% Fe_2O_3, 8–10 wt% Cr_2O_3, Al_2O_3
[c]Medium temp. shift catalyst (ICI Katalco): typical composition: 57 wt% CuO, 31 wt% ZnO, 12 wt% Al_2O_3
[d]0.5 wt% Pt/Al_2O_3 (Johnson Matthey)
[e]1 wt% Pt/SiO_2 (Sigma Aldrich)
[f]1 wt% Pt/CeO_2, (In-house synthesised catalyst, Pt content measured by ICP)
[g]1 wt% Pt/20 wt% CeO_2/SBA15, (In-house synthesised catalyst, Pt content measured by ICP)
[h]1 wt% Pt/0.7:0.3 CeO_2-ZrO_2, (In-house synthesised catalyst, Pt content measured by ICP)

synthesised catalysts are given in Table 8.3. Additionally, the nominal compositions according to the datasheets of the suppliers or measured by ICP are included.

8.3.2 *Catalytic Tests*

The experimental setup and the procedure for the catalyst screening tests have been described in detail in Chap. 3, Sect. 3.3.1. In brief, 30 mg of catalyst powder were loaded into a continuous microreactor and activated in situ flowing 40 mL·min^{-1} of a mixture 10 % H_2 in N_2 at 573 K for 4 h. Subsequently, the reactor was cooled down to 373 K and the reactants were fed. The temperature was increased at a rate of 10 K min^{-1} while the composition of the effluent was monitored online using a mass spectrometer. The experiments were conducted at 1 bar.

The reverse water gas shift activities and selectivites of the different catalysts tested are given in Table 8.4. In all cases, CO, CH_4 and H_2O were the only products detected between 373 and 773 K. No methanol was formed under the operating conditions used. The equilibrium conversions to CO increase with the reaction temperature due to the endothermicity of the transformation. Commercial Co/Mo based catalyst exhibited the lowest conversions among the samples tested. The selectivity to CO was also found to be low at 573 and 673 K but increases at 773 K due to thermodynamic limitations on the methanation reaction. The poor performance of Co/Mo is related to the absence of sulphur-containing molecules in the

Table 8.4 Temperature-programmed RWGS. CO_2 conversion and carbon selectivities, CO and CH_4 selectivities. Conditions: CO_2:H_2 = 1:1, GHSV = 60,299 h^{-1}, 1 bar

Catalysts	573 K[a]			673 K[b]			773 K[c]		
	x	S_{CO}	S_{CH_4}	x	S_{CO}	S_{CH_4}	x	S_{CO}	S_{CH_4}
Co/Mo	0.00	0.33	0.67	0.02	0.55	0.45	0.04	0.98	0.02
Fe/Cr	0.04	0.83	0.17	0.13	0.99	0.01	0.26	0.99	0.01
Cu/ZnO	0.12	0.98	0.02	0.23	0.99	0.01	0.31	1.00	0.00
Pt/Al_2O_3	0.03	0.64	0.36	0.03	0.85	0.15	0.08	0.96	0.04
Pt/SiO_2	0.01	0.64	0.36	0.02	0.95	0.05	0.08	0.98	0.02
Pt/CeO_2	0.05	0.96	0.04	0.16	0.97	0.03	0.25	0.89	0.11
Pt/CeO_2/SBA15	0.06	0.96	0.04	0.18	0.97	0.03	0.28	0.89	0.11
Pt/CeO_2/ZrO_2	0.07	0.97	0.03	0.20	0.96	0.04	0.29	0.84	0.16

[a]Equilibrium conversion at 573 K = 0.14
[b]Equilibrium conversion at 673 K = 0.23
[c]Equilibrium conversion at 773 K = 0.31

surrounding gas, which are required in low concentrations to activate the catalyst sites [9]. Significantly better performances were observed with the commercial high temperature (Fe/Cr) and low temperature (Cu/ZnO) shift catalysts. As expected, the CO_2 conversions obtained with the Fe/Cr catalyst increase with the reaction temperature while methane formation decreases. Cu/ZnO is significantly more active than Fe/Cr, showing conversions close to equilibrium and a negligible formation of methane. The results indicate that in the temperature window presented in Table 8.4, the reverse water gas shift is thermodynamically limited over the Cu/ZnO catalyst which is desirable for the sorption- enhanced methanol-to-shift process. The activity and the selectivity of platinum supported on alumina or silica are relatively low compared to Cu/ZnO but both catalyst become remarkably selective to CO at high temperatures. Below 673 K the activity and selectivity of Pt was found to increase upon addition of CeO_2, in agreement with previous reports [8, 12]. Methane formation over the three Pt/CeO_2 systems increases gradually with temperature but decreases above 773 K due to thermodynamic limitations. The conversions obtained over the Pt/CeO_2 based catalysts seem to be related to the surface area of the materials, Table 8.3, but strong metal-support interactions cannot be ruled out.

The carbon conversions, hydrogen yields and carbon selectivities obtained during temperature-programmed methanol decomposition are given in Tables 8.5 and 8.6. A blank experiment showed that the thermal decomposition of methanol starts around 723 K to give hydrogen and carbon monoxide as main gaseous products. Co/Mo was found to be active in the decomposition of methanol but relatively high amounts of methane and CO_2 were detected. Molybdenum sulphide is known to catalyse the synthesis of methanol producing methane as secondary product unless it is promoted with alkali metals [13]. Part of the water formed during the hydrogenation of CO to methane is consumed by water gas shift which

Table 8.5 Temperature-programmed methanol decomposition conversion and H_2 yield. Conditions: $p_{CH_3OH} = 50$ mbar, GHSV = 83,060 h^{-1}, 1 bar

Catalysts	573 K[a]		673 K[b]		773 K[c]	
	x	Y_{H_2}	x	Y_{H_2}	x	Y_{H_2}
Co/Mo	0.64	0.01	0.86	0.08	0.99	0.20
Fe/Cr[d]	0.13	0.14	0.97	0.60	1.00	0.58
Cu/ZnO[d]	0.65	0.53	0.93	0.68	0.93	0.65
Pt/Al_2O_3	0.03	0.04	0.47	0.39	0.90	0.64
Pt/SiO_2	0.15	0.14	0.90	0.69	1.00	0.73
Pt/CeO_2	0.58	0.43	0.97	0.63	0.99	0.53
Pt/CeO_2/SBA15	0.75	0.56	0.99	0.65	1.00	0.51
Pt/CeO_2/ZrO_2	0.66	0.55	1.00	0.67	1.00	0.45

[a]Equilibrium conversion at 573 K = 0.99
[b]Equilibrium conversion at 673 K = 1.00
[c]Equilibrium conversion at 773 K = 1.00
[d]The discrepancy between the carbon and oxygen element mass balances is around 20 %

Table 8.6 Temperature-programmed methanol decomposition. CO, CH_4 and CO_2 carbon selectivities. Conditions: $p_{CH_3OH} = 50$ mbar, GHSV = 83,060 h^{-1}, 1 bar

Catalysts	573 K			673 K			773 K		
	S_{CO}	S_{CH_4}	S_{CO_2}	S_{CO}	S_{CH_4}	S_{CO_2}	S_{CO}	S_{CH_4}	S_{CO_2}
Co/Mo	0.82	0.13	0.05	0.65	0.22	0.13	0.54	0.29	0.17
Fe/Cr	0.42	0.10	0.48	0.34	0.23	0.43	0.40	0.22	0.38
Cu/Z_nO	0.77	0.01	0.22	0.79	0.03	0.18	0.65	0.03	0.31
Pt/Al_2O_3	0.85	0.13	0.00	0.97	0.02	0.01	0.96	0.01	0.02
Pt/SiO_2	0.93	0.05	0.02	0.99	0.01	0.01	0.96	0.01	0.04
Pt/CeO_2	0.94	0.01	0.05	0.79	0.05	0.15	0.53	0.20	0.27
Pt/CeO_2/SBA15	0.96	0.00	0.04	0.79	0.06	0.15	0.48	0.24	0.28
Pt/CeO_2/ZrO_2	0.91	0.01	0.07	0.62	0.13	0.25	0.28	0.33	0.39

produces CO_2. The low temperature Cu/ZnO shift catalyst exhibits high conversions of methanol to CO and H_2 between 573 and 773 K with a significant production of CO_2. Different intermediates and by-products of methanol decomposition on Cu/ZnO such as methoxy species, formaldehyde, methyl formate and dimethyl ether can be oxidised to produce CO_2 [7, 14]. However, none of these species was detected by the online MS analyser. Moreover, the source of the oxidising agent is intriguing since water was not co-fed and low amounts of water are expected to be formed by methanation or other parallel reactions. Disproportionation of CO to form solid carbon and CO_2 (Boudouard reaction) is known to occur under the reducing conditions that prevail in the synthesis or decomposition of methanol [15]. Alternatively, above 573 K and in the presence of

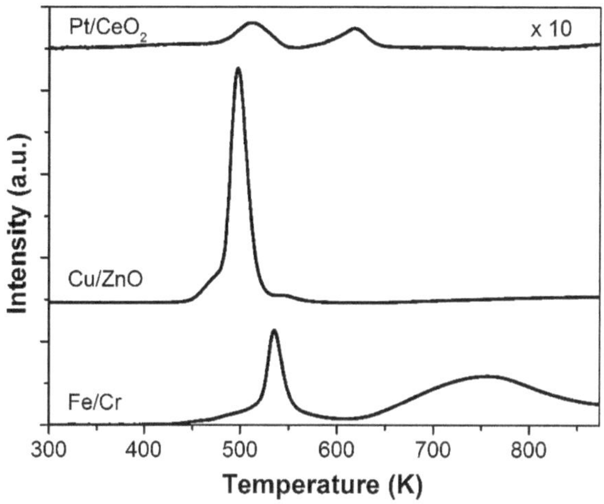

Fig. 8.10 Temperature-programmed reduction profiles for Fe/Cr, Cu/ZnO and Pt/CeO_2 catalysts

carbon monoxide and hydrogen, ZnO can reduce and form an α-brass alloy with Cu producing CO_2 and H_2O [16, 17]. The Boudouard reaction and brass formation are expected to eventually deactivate the Cu/ZnO catalyst but such effect cannot be appreciated in temperature-programmed reaction experiments. The water co-fed during the methanol-to-shift process is expected to mitigate the deactivation by carbon deposition and brass formation.

The high temperature shift Fe/Cr catalyst is slightly less active than Cu/ZnO and produces more CO_2. Iron is known to catalyse the deposition of graphene layers from CO, giving CO_2 as a side product [18]. Additionally, further reduction of the Fe/Cr system by CO during the screening test is likely to occur since high temperature water gas shift is known to proceed via a redox mechanism between 573 and 773 K [9]. This is in agreement with the temperature-programmed reduction (TPR) experiments shown in Fig. 8.10. The TPR analyses were conducted in a quartz tube microreactor (d_t = 6.35 mm, wall thickness = 1.21 mm, length = 270 mm) heated by a concentric furnace. A gas mixture 10 % H_2 in N_2 was fed with a flow rate of 50 mL min^{-1} and the temperature was increased from 298 to 873 K at 10 K min^{-1} (see Sect. 3.4.8, Chap. 3).

Pt/Al_2O_3 and Pt/SiO_2 are inactive at 573 K but high conversions and remarkable selectivities to CO are observed at 673 and 773 K, Tables 8.5 and 8.6. The negligible formation of CO_2 is related to the non-reducibility of the supports under the conditions tested. In the presence of cerium oxide the activity of Pt at low temperatures is enhanced significantly while CO_2 formation increases very modestly. The selectivities and hydrogen yields observed are similar to values reported by other authors [8]. The TPR profile of Pt/CeO_2 reveals a very small consumption of H_2 implying that bulk reduction of the support does not take place. Therefore, the low levels of CO_2 detected with the Pt/CeO_2-based samples are likely to be associated to the removal of adsorbed oxygen and to the creation of surface vacancies [12, 19].

From the screening tests Cu/ZnO and Pt/CeO_2 were identified as the most promising candidates to catalyse the simultaneous methanol decomposition and water gas shift in the temperature range in which LDOs show adequate capacities (573–673 K). In this range of temperature, the catalysts showed higher activity and selectivity for the RWGS and methanol decomposition reactions compared to the rest of the catalysts screened. Consequently, the studies presented hereafter focus on these two systems.

8.3.3 Water Gas Shift Over Cu/ZnO and Pt/CeO_2 Catalysts

Temperature-programmed water gas shift experiments were carried out over the Cu/ZnO and Pt/CeO_2 catalysts at 0.1 MPa. Carbon monoxide and water were fed in a molar ratio 1 to 2 with a gas hourly space velocity (GHSV) of 178,000 h^{-1}. The temperature was increased from 373 to 873 K at 10 K min^{-1}. The conversion trends are presented in Fig. 8.11 where the equilibrium values are also included.

At low temperature Cu/ZnO is more active than Pt/CeO_2 but the trend is inverted around 680 K. With the space velocity used, the conversion of Cu/ZnO is always below equilibrium while Pt/CeO_2 reaches the thermodynamic limit at 770 K and above. CO_2 was the main carbon-containing molecule in the gas phase with selectivities close to 100 %. In the case of Cu/ZnO the methane concentration was below the detection limit of the analyser while very low levels (maximum of 0.1 % at 773 K) were measured over Pt/CeO_2.

The stability of the catalysts was assessed in WGS experiments with duration of 24 h conducted at 673 K feeding H_2O and CO in a ratio 2:1. The activity of both

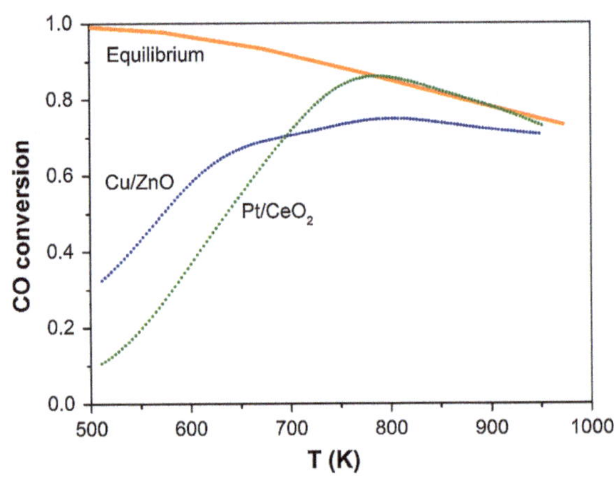

Fig. 8.11 Temperature-programmed WGS over Cu/ZnO and Pt/CeO_2 catalysts

catalysts remained constant throughout the tests with carbon conversions of 74.2 and 51.1 % for Cu/ZnO and Pt/CeO_2 respectively. The equilibrium conversion at the conditions of the tests is 93.3 %. The selectivities to CO_2 were 99.9 % for Cu/ZnO and 99.8 % for Pt/CeO_2, with methane as the only by-product. These values were corroborated by offline GC analysis of gas samples taken at different times on stream.

8.4 Concluding Remarks

Among the various possible applications that the hydrotalcite derivatives presented in the preceding chapters may find, sorption-assisted hydrogen production stands out as an interesting option for process intensification. In this regard the water gas shift reaction has been widely studied, but other schemes also worth attention. The thermodynamic analysis of the sorption-enhanced methanol-to-shift (SEMTS) revealed that in the temperature range between 573 and 773 K, the removal of 90 % of the CO_2 formed during the transformation results in significant increments in the hydrogen yield. This temperature window coincides with the conditions where LDH derivatives are adequate adsorbents. It was found that the increase in hydrogen productivity for SEMTS stems from the conversion of the H_2 and CO contained in the methanol molecules. Therefore, the main advantage of SEMTS is that the combination of the endothermic methanol decomposition with the exothermic WGS and CO_2 adsorption allows, in principle, for autothermal operation which is not possible with SEWGS alone. A remarkable finding from the thermodynamic studies is that the combined water gas shift and alcohol decomposition is efficient only if the formation of methane is prevented. Consequently, the sorption-enhanced methanol-to-shift process requires very selective catalysts.

Catalyst screening tests showed that the commercial $Cu/Z_nO/Al_2O_3$ catalyst for mid-temperature water gas shift and the in-house synthesised Pt/CeO_2-based catalysts are promising candidates for the methanol-to-shift process enhanced by CO_2 adsorption over LDO derivatives. Both systems were found to be active for water gas shift and methanol decomposition and exhibited outstanding selectivities with negligible methane formation. Stability tests under water gas shift conditions showed constant selectivities above 98 % with no decrease in the conversion levels during experiments lasting 24 h.

As indicated in the introduction to this chapter, the studies here included aim to serve as a guideline for future work on methanol-to-shift and, more generally, on alcohol-to-shift processes enhanced by adsorption. Detailed analysis of the catalytic performance under the actual methanol-to-shift feedstock and operating conditions is still required. Likewise, it is necessary to prove experimentally the viability of the sorption-enhanced methanol-to-shift process using LDO derivatives as adsorbents. In particular, it is crucial to assess the reactivity and stability of unsupported and supported LDOs in the presence of reactants and products.

References

1. Comas, J., Laborde, M., & Amadeo, N. (2004). Thermodynamic analysis of hydrogen production from ethanol using CaO as a CO_2 sorbent. *Journal of Power Sources, 138*(1–2), 61–67.
2. Chen, H., Zhang, T., Dou, B., Dupont, V., Williams, P., Ghadiri, M., & Ding, Y. (2009). Thermodynamic analyses of adsorption-enhanced steam reforming of glycerol for hydrogen production. *International Journal of Hydrogen Energy, 34*(17), 7208–7222.
3. Nielsen, P. E. H., Hansen, J. B., & Schiødt, N. C. (2009). Process for the preparation of a hydrogen-rich stream. US 7,527,781 B2, 2009.
4. Smith, J. M., Van Ness, H. C., & Abbott, M. M. (2001). *Introduction to chemical engineering thermodynamics*. London, UK: McGraw-Hill.
5. Ding, Y., & Alpay, E. (2000). Equilibria and kinetics of CO_2 adsorption on hydrotalcite adsorbent. *Chemical Engineering Science, 55*(17), 3461–3474.
6. Lwin, Y., Daud, W. R. W., Mohamad, A. B., & Yaakob, Z. (2000). Hydrogen production from steam–methanol reforming: thermodynamic analysis. *International Journal of Hydrogen Energy, 25*(1), 47–53.
7. Choi, Y., & Stenger, H. (2002). Kinetics of methanol decomposition and water gas shift reaction on a commercial Cu-ZnO/Al_2O_3 catalyst. *Fuel Chemistry Division Preprints, 47*, 723–724.
8. Brown, J. C., & Gulari, E. (2004). Hydrogen production from methanol decomposition over Pt/Al_2O3 and ceria promoted Pt/Al_2O_3 catalysts. *Catalysis Communications, 5*(8), 431–436.
9. Hinrichsen, K.-O., Kochloefl, K., & Muhler, M. (2008). Water gas shift and COS removal. In *Handbook of heterogeneous catalysis*. Germany: Wiley-VCH Verlag GmbH & Co. KGaA.
10. van Dijk, H. A. J., Walspurger, S., Cobden, P. D., Jansen, D., van den Brink, R. W., & de Vos, F. G. (2009). Performance of water-gas shift catalysts under sorption-enhanced water-gas shift conditions. *Energy Procedia, 1*(1), 639–646.
11. Sá, S., Silva, H., Brandão, L., Sousa, J. M., & Mendes, A. (2010). Catalysts for methanol steam reforming—a review. *Applied Catalysis, B: Environmental, 99*(1–2), 43–57.
12. Yeung, C. M. Y., & Tsang, S. C. (2009). Noble metal core—ceria shell catalysts for water—gas shift reaction. *The Journal of Physical Chemistry C, 113*(15), 6074–6087.
13. Hansen, J. B., & Højlund Nielsen, P. E. (2008). Methanol synthesis. In *Handbook of heterogeneous catalysis*. Wiley-VCH Verlag GmbH & Co. KGaA.
14. Huffman, J. R., & Dodge, B. F. (1929). Decomposition of methanol over catalysts composed of oxides of zinc and chromium. *Industrial and Engineering Chemistry, 21*(11), 1056–1061.
15. Twigg, M., & Spencer, M. (2003). Deactivation of copper metal catalysts for methanol decomposition, methanol steam reforming and methanol synthesis. *Topics in Catalysis, 22*(3–4), 191–203.
16. Spencer, M. S. (1987). α-Brass formation in copper/zinc oxide catalysts: II. Diffusion of zinc in copper and α-brass under reaction conditions. *Surface Science, 192*(2–3), 329–335.
17. Nakamura, J., Nakamura, I., Uchijima, T., Watanabe, T., & Fujitani, T. (1996). Model studies of methanol synthesis on copper catalysts. In W. Joe, W. N. D. E. I. Hightower & T. B. Alexis (Eds.), *Studies in surface science and catalysis* (Vol. 101, pp. 1389–1399). Amsterdam: Elsevier.
18. Rakov, E. G. (2006). *Nanotubes and nanofibers*. FL, USA: Taylor and Francis Group.
19. Yeung, C., & Tsang, S. (2009). Microemulsion prepared pt in ceria: Catalytically active for water gas shift reaction but totally inert for methanation. *Catalysis Letters, 128*(3–4), 349–355.

Chapter 9
Conclusions and Future Work

Sorption-enhanced technologies for hydrogen production require the development of adsorbents with adequate CO_2 capacities and multicycle stability that can operate in the region in which the reactions are catalytically feasible but thermodynamically limited. In this regard, various studies indicate that layered double hydroxides (LDHs) are promising materials to conduct the in situ removal of carbon dioxide in the water gas shift stage since they are easy to regenerate and are highly selective towards CO_2 at temperatures between 573 and 773 K. Previous research efforts have focused primarily on adding promoters to increase the adsorption capacity of LDHs but their impact on the stability has been often disregarded. An alternative approach is the use of supporting materials to increase the intrinsic capacities (per mass of LDH) which renders stable adsorbents. However, due to the large amounts of support often used, the few materials reported so far exhibit low CO_2 uptakes per volume of adsorbent making them unattractive for industrial applications. In this thesis the CO_2 adsorption characteristics of novel LDHs supported on multi-walled carbon nanotubes (MWCNTs) and graphene oxide (GO) have been studied with emphasis on the range of low support loadings. It is the first time that the adsorption properties of this type of composites are investigated. One of the main aims of this research was to contribute to the understanding of the equilibria and kinetics of CO_2 adsorption on these materials. To this end the physicochemical properties of the adsorbents were studied in detail by various characterisation techniques and were correlated with their adsorption capacity and multicycle stability. A systematic study of pure LDHs and carbon containing hybrids was carried out to clarify the role of the support on the CO_2 adsorption. Additionally, the promotion of the materials by alkali ions was investigated providing valuable knowledge in this area.

The coprecipitation of aluminium and magnesium nitrates onto well-dispersed oxidised MWCNTs was found to be an easy preparation method that resulted in a good interaction between the LDH and the carbon support. Microscope images showed that all the LDH particles were attached to MWCNTs indicating that the

D. Iruretagoyena Ferrer, *Supported Layered Double Hydroxides as CO_2 Adsorbents for Sorption-enhanced H_2 Production*, Springer Theses,
DOI 10.1007/978-3-319-41276-4_9

association between the positively-charged LDH sheets and the negatively charged oxidised nanotubes successfully leads to the deposition of LDH on the support. The multi-walled carbon nanotubes were shown to be inert for the adsorption of CO_2. As the loading of support in the hybrids increases, the CO_2 uptake per mass of LDH increases due to the creation of crystalline defects (stacking faults were detected by XRD) and to the enhancement of the gas accessibility provided by the MWCNT framework. The presence of even moderate loadings of carbon nanotubes resulted in significantly more stable materials. This remarkable finding indicates that, although the unsupported LDH exhibits an initial capacity per volume of adsorbent that is higher than any of the nanotube hybrids, after consecutive adsorption-desorption cycles the trend is inverted due to an improved stability, resulting in the existence of an optimum content of MWCNTs (35–50 wt%).

The direct precipitation of LDH nanoparticles onto GO resulted in the intercalation of LDH platelets and GO nanosheets when the support loadings are below 50 wt% as revealed by TEM images and X-ray diffractograms. Higher GO contents favour restacking of graphene layers rendering materials with very low surface areas. As in the case of MWCNTs, the adsorption of CO_2 on GO was found to be negligible. There is an increase in the intrinsic adsorption capacity of LDHs supported on GO but the improvement is minor compared to the observed when nanotubes are used. On the other hand, the adsorbents were found to become remarkably stable with GO loadings as low as 5 wt%, which are notably lower than the loadings of nanotube required for similar enhancements. The effect is attributable to the compatibility of charge and geometry between the GO layers and the LDH platelets and therefore significantly less GO than MWCNTs is required to stabilise the same amount LDH. This results in exceptionally stable LDH/GO hybrids with high CO_2 capacities per total volume of adsorbent which are promising for industrial applications.

The use of graphene oxide as stabiliser for the LDH platelets does not modify significantly the nature of the chemisorption sites as revealed by temperature-programmed desorption experiments, which is in agreement with the minor increase in the CO_2 intrinsic capacity upon GO addition. A comparison between fresh materials and samples spent in multiple adsorption-desorption cycles revealed that the presence of GO helps to maintain the heterogeneity of the adsorption sites by mediation of the thermally driven crystallite growth. Accordingly, while the first-contact adsorption equilibrium of the materials was found to be described by the Freundlich model, which is characteristic of heterogeneous surfaces, after temperature-swing cycles the isotherms tend to fit a langmuirian expression as the surface becomes more uniform. Increasing amounts of GO help to maintain the Freundlich profile for a larger number of cycles. Moreover, it was found that the Toth isotherm can be used to describe simultaneously the loss of CO_2 capacity and the transition of the surface. These findings constitute an important contribution to the understanding of the deactivation of LDO adsorbents which is often disregarded in the specialised literature. The magnitudes of the isosteric heat of adsorption were estimated to be between 25 and 50 kJ mol^{-1} suggesting that physisorption contributes to the adsorption capacities of the materials [1].

The addition of sodium, potassium and caesium ions increases the CO_2 uptake by more than 40 % without affecting the stability of unsupported and GO-supported LDHs to any significant extent. Comparative studies have been reported previously using LDHs with the same content of alkali ions in a mass basis, resulting in marked differences in the number of promoting species loaded. Conversely, in the present work the samples were prepared aiming to obtain the same amount of alkali species in a molar basis. Alkali promoters increase the density of strong basic sites which correlates with the enhancement of capacity observed at 573 K in adsorption-desorption cycles. The effectiveness of the promotion seems to increase with the basicity of the alkali ions and therefore the presence of potassium was found to be more beneficial than sodium. However, in spite of the more basic character of caesium, its tendency to form bulky agglomerates hinders the effective interaction between Cs and LDH. The efficiency of the promotion by impregnation with aqueous Na_2CO_3 or by leaving sodium residues from the basic solution used for the coprecipitation revealed that the latter method results in higher CO_2 capacities. Therefore, it is evident that the promotion requires an efficient interaction between the alkali and the LDH platelets.

Breakthrough experiments conducted in a fixed-bed column revealed a fast initial CO_2 adsorption followed by a more gradual uptake as the inlet concentration is approached. The experimental data can be successfully modelled by the linear driving force approximation and the Toth isotherm, although the description may be improved if limitations to the transport of energy are considered. The magnitude of the mass transfer coefficients indicates that the presence of GO as support does not affect nor enhance the CO_2 adsorption kinetics of LDH.

While temperature-swing experiments showed a gradual and continuous decay in the CO_2 capacity, regeneration by isothermal N_2 purge in the fixed-bed column showed that the loss of capacity occurs mainly in the first cycle possibly due to irreversible adsorption. This indicates that thermally driven effects are likely to contribute to the deactivation during temperature-swing operation. The addition of 16 vol. % water to the feed increases the stability of the materials in temperature-swing cycles and enhances the CO_2 capacities by 20 % due to the rehydration of the LDH surface. No gasification of the support was observed during the wet experiments.

LDH derivatives can operate adequately under the conditions required for the sorption enhancement of the water gas shift reaction. Besides this application, a thermodynamic analysis of the simultaneous water gas shift and methanol decomposition (methanol-to-shift) demonstrated the convenience of improving the process by the in situ removal of CO_2 using LDHs. Although methanol-to-shift has been reported previously in the literature, it is the first time that sorption enhancement of this technology is proposed as a strategy to increase the hydrogen yield while reducing the energy demand. Additionally, $Cu/ZnO/Al_2O_3$ and Pt/CeO_2-based catalysts were identified to be promising systems to catalyse the sorption- enhanced water gas shift and the sorption-enhanced methanol-to-shift.

So far there has been a gap in the knowledge of suitable adsorbents that operate under the conditions of water gas shift and methanol-to-shift, i.e. between 473 and

773 K. High capacity materials for CO_2 sequestration at low temperature (below 350 K) such as zeolites, activated carbons and metal-organic frameworks are well-known and are even commercially available, but their capacities at higher temperatures and in the presence of water are relatively poor (see Sect. 2.3, Chap. 2) [2]. Other options such as liquid and solid-supported amine absorbers are not feasible either since amines decompose around 373 K. Likewise, high temperature chemisorbents such as lithium zirconates and calcium oxides exhibit very slow kinetics and fast deactivation profiles, and they require large amounts of energy to be regenerated. Therefore, despite their relatively low adsorption capacities, LDH based materials appear as the most adequate candidates for application in sorption-enhanced shift processes and pilot projects are being undertaken to prove the concept [3]. In this context the supported materials developed in this research are remarkably stable compared to unsupported LDHs and exhibit satisfactory CO_2 capacities per volume of total adsorbent. Compared to other supports, the CO_2 capacity of LDH/MWCNTs hybrids (per volume of adsorbent) is twice the capacity of LDH supported on CNFs. In addition, although the first-contact capacity of MWCNTs materials was found to be similar to that of a commercial LDH/alumina adsorbent, the carbon supported LDHs were markedly more stable resulting in higher capacities after multicycle experiments. On the other hand, GO containing hybrids exhibit the highest capacity per volume of adsorbent reported for supported LDHs so far, being 50 % higher than that of LDHs supported on MWCNTs or alumina. Consequently, the findings of this work constitute an important step towards the industrial implementation of sorption-enhanced shift reactions.

9.1 Future Work

Although the present research was conducted to improve the competitiveness of layered double hydroxides as solid adsorbents, the outstanding stability and the tuneable acid-base properties of these materials make them attractive for base-catalysed reactions. In this regard, the thorough characterisation of the basic sites in the surface of unsupported, GO-supported and alkali-promoted LDHs may be useful as guideline for future catalysis-oriented studies. The reactions in which these materials can be used are normally conducted at low temperatures and therefore the thermal decomposition of the carbon support is unlikely. However, the reactivity of carboxylic groups and base impurities that may remain attached to the surface of the supports needs to be investigated.

While the present work sets the foundations for sorption enhanced water gas shift and methanol-to-shift processes using carbon supported LDHs, research efforts are still necessary to prove the viability of these technologies. As discussed in Chap. 5, graphene oxide seems to be a more effective support than MWCNTs for LDHs resulting in higher adsorption capacities per volume of total adsorbent. To date the commercial providers for graphene oxide are scarce and the production is costly (around £250 per gram), so the composite materials here presented would hardly find an immediate application. However, research efforts are being devoted to improve current production methods for graphene and graphene oxide. This is expected to lead to a price decrease similar to the observed for the production of carbon nanofibers and nanotubes in the last decades. Currently, carbon nanofibers and carbon nanotubes would be the most likely candidates to be used as a support since the technologies for dispersion and large scale production of the materials are already available [4].

Besides addressing the challenge of scaling up the production of the hybrids, further fundamental understanding of the adsorbents is needed, particularly regarding the nature of the promotion of the performance of the materials in the presence of water. In relation to sorption-enhanced applications, it is necessary to carry out water gas shift and methanol-to- shift experiments under the pressures and gas compositions that are relevant for large scale processes. Particular attention deserves the investigation of the influence of the regeneration conditions on the stability and the performance of catalysts and adsorbents in multicycle experiments that approach the number of cycles expected for an industrial operation. Likewise the mechanical strength of the adsorbents needs to be assessed and possibly improved by the use of binders such as alumina.

References

1. Ding, Y., & Alpay, E. (2000). Equilibria and kinetics of CO_2 adsorption on hydrotalcite adsorbent. *Chemical Engineering Science, 55*(17), 3461–3474.
2. Choi, S., Drese, J. H., & Jones, C. W. (2009). Adsorbent materials for carbon dioxide capture from large anthropogenic point sources. *ChemSusChem, 2*(9), 796–854.
3. ECN develops enhanced SEWGS technology. (2012). *Membrane Technology, 2012*(9), 7.
4. Wong, H. S. P., & Akinwande, D. (2011). *Carbon nanotube and graphene device physics*. NY, USA: Cambridge University Press.

Appendix A
Literature Review

A.1 Theoretical Justification of the Freundlich Isotherm (Sect. 2.2.3.2)

In the theoretical derivation of the Freundlich isotherm, the adsorbent surface is divided in a group of domains which can be described by the Langmuir equation [1, 3]. The domain distribution is assumed to be exponentially dependent on the heat of adsorption, therefore the equilibrium surface coverage is expressed as

$$\theta = \int \alpha(Q_{ads})\theta(Q_{ads})dQ_{ads} = \int \left(\alpha_o \mathrm{e}^{-Q_{ads}/Q_{adsm}}\right)\left(\frac{bp_A}{1+bp_A}\right)dQ_{ads} \tag{A.1}$$

Including the dependence of b on Q_{ads} and reordering the expression

$$\theta = \int_{-\infty}^{+\infty} \frac{\alpha_o \mathrm{e}^{-Q_{ads}/Q_{adsm}}}{1 + \frac{\mathrm{e}^{-Q_{ads}/RT}}{b_o p_A}} dQ_{ads} \tag{A.2}$$

To solve the integral, the equation is written as

$$\theta = c \int_{-\infty}^{+\infty} \frac{\mathrm{e}^{-x/x_m}}{1 + \frac{a}{b}\mathrm{e}^{-x/RT}} dx \tag{A.3}$$

where $c = \alpha_o$, $a = 1$, $b = b_o p_A$, $x = Q_{ads}$, and $x_m = x_m$. A new variable, Z is defined as

$$Z = \frac{a}{b}\mathrm{e}^{-x/RT} \quad \therefore \quad dZ = \left(-\frac{1}{RT}\right)\left(\frac{a}{b}\right)\mathrm{e}^{-x/RT}dx \tag{A.4}$$

D. Iruretagoyena Ferrer, *Supported Layered Double Hydroxides as CO_2 Adsorbents for Sorption-enhanced H_2 Production*, Springer Theses,
DOI 10.1007/978-3-319-41276-4

Changing the integration variable and limits

$$\theta = c \int_{-\infty}^{+\infty} \frac{e^{-x/x_m}}{1 + \frac{a}{b}e^{-x/RT}} \cdot \frac{\left(-\frac{1}{RT}\right)\left(\frac{a}{b}\right)e^{-x/RT}}{\left(-\frac{1}{RT}\right)\left(\frac{a}{b}\right)e^{-x/RT}} dx \tag{A.5}$$

$$\theta = c \int_{\infty}^{0} \frac{e^{-x/x_m}}{1+Z} \cdot (-RT)\left(\frac{a}{b}\right)^{-1} e^{x/RT} dZ \tag{A.6}$$

$$\theta = cRT\left(\frac{a}{b}\right)^{-1} \int_{0}^{\infty} \frac{e^{[(-x/x_m)(RT/x_m-1)]}}{1+Z} dZ \tag{A.7}$$

$$\theta = cRT\left(\frac{a}{b}\right)^{-1} \int_{0}^{\infty} \frac{\left(\frac{a}{b}\right)^{(RT/x_m-1)}\left[e^{(-x/x_m)}\right]^{(RT/x_m-1)}}{1+Z} \left(\frac{a}{b}\right)^{(1-RT/x_m)} dZ \tag{A.8}$$

$$\theta = cRT\left(\frac{a}{b}\right)^{-RT/x_m} \int_{0}^{\infty} \frac{Z^{(RT/x_m-1)}}{1+Z} dZ \tag{A.9}$$

The integral $\int_0^{\infty} \frac{Z^{(RT/x_m-1)}}{1+Z} dZ$ has the value $\pi \operatorname{cosec}(\pi RT/x_m)$ and does not contain any term involving the partial pressure of the gas. Consequently, the infinite integral can be grouped within the constant c so

$$\theta = cRT(b)^{-RT/x_m} \tag{A.10}$$

Replacing the original variables in the resulting equation, a form of the Freundlich isotherm is obtained:

$$\theta = \alpha_o RT(b_o p_A)^{RT/Q_{adsm}} \tag{A.11}$$

Appendix B
Experimental Methods

B.1 Adsorption Capacity Calculation in a Fixed-Bed (Sect. 3.2.2.3)

The mass balance of CO_2 in a packed bed adsorption column in which the radial gradients are negligible can be written as:

$$\varepsilon_T \frac{\partial C_{CO_2}}{\partial t} + \frac{\partial (u C_{CO_2})}{\partial z} = \varepsilon_b \frac{\partial}{\partial z}\left(D_{ax}\frac{\partial C_{CO_2}}{\partial z}\right) - \rho_b \frac{\partial q_{CO_2}}{\partial t} \tag{B.1}$$

ε_b and ε_T are the interparticle bed and the total (inter- and intraparticle) bed porosities respectively, ρ_b is the bed density of the adsorbent, D_{ax} is the axial dispersion coefficient, u is the superficial gas velocity, q are the moles of CO_2 retained per mass of adsorbent, and C_{CO_2} is the concentration of CO_2 in the gas phase.

If no axial dispersion occurs in the gas front and the balance is written in terms of the column volume, the following expression is obtained:

$$\frac{\partial}{\partial V}(Q C_{CO_2}) = -\varepsilon_T \frac{\partial C_{CO_2}}{\partial t} - \rho_b \frac{\partial q_{CO_2}}{\partial t} \tag{B.2}$$

where Q is the volumetric gas flow rate. Integrating over the total bed volume and considering that at $V = 0$ the concentration and the flow rate are those of the feed, i.e. $Q C_{CO_2} = Q_f C_{CO_2 f}$, the mass balance can be written as

$$Q C_{CO_2} = -\varepsilon_T V \frac{\partial C_{CO_2}}{\partial t} - \rho_b V \frac{\partial q_{CO_2}}{\partial t} + Q_f C_{CO_2 f} \tag{B.3}$$

Reordering the expression and integrating from $t = 0$ (step input) to $t = t^{\circ}$, which corresponds to the time at which C_{CO_2} in the outlet equals $C_{CO_2 f}$

D. Iruretagoyena Ferrer, *Supported Layered Double Hydroxides as CO_2 Adsorbents for Sorption-enhanced H_2 Production*, Springer Theses,
DOI 10.1007/978-3-319-41276-4

$$m_a \int_0^q \partial q = \int_0^{t^\circ} Q_f C_{CO_2f} \partial t - \int_0^{t^\circ} Q C_{CO_2} \partial t - \int_0^{t^\circ} \varepsilon_T V \frac{\partial C_{CO_2}}{\partial t} \partial t \quad \text{(B.4)}$$

m_a is the mass of adsorbent in the column. The integral results in

$$m_a q_{CO_2} = Q_f C_{CO_2f} t^\circ - \int_0^{t^\circ} Q C_{CO_2} \partial t - \varepsilon_T V C_{CO_2f} \quad \text{(B.5)}$$

Considering that the change in the flow rate throughout the packed column is due to the adsorption of CO_2 molecules exclusively (i.e. $Q_f C_{Arf} = Q C_{Ar}$) and assuming that the adsorptive is an ideal gas:

$$q_{CO_2} = \frac{Q_f}{m_a}\left[t^\circ C_{CO_2f} - \int_0^{t^\circ} \frac{C_{Arf} C_{CO_2}}{C_{Ar}} dt\right] - \frac{\varepsilon_T V_b P_b y_{CO_2f}}{R T_b m_a} \quad \text{(B.6)}$$

$$q_{CO_2} = \frac{Q_f P_s}{R T_s m_a}\left[t^\circ y_{CO_2f} - y_{Ar_f} \int_0^{t^\circ} \frac{y_{CO_2}}{y_{Ar}} dt\right] - \frac{\varepsilon_T V_b P_b y_{CO_2f}}{R T_b m_a} \quad \text{(B.7)}$$

B.2 Parameters Used in the Calculation of the Mass Transfer Biot Number (Sect. 3.2.2.4)

Molar composition	u (m s^{-1})	μ_{gas}^{a} (kg m^{-1} s^{-1})	ρ_{gas}^{b} (kg m^{-3})	Dm^{c} (m^2 s^{-1})	Dp^{d} (m^2 s^{-1})	Ds^{e} (m^2 s^{-1})
2 % CO_2, 98 % Ar	0.03	4.2E−05	8.8E−01	4.9E−05	7.4E−07	1.5E−09
2 % CO_2, 98 % Ar	0.09	4.2E−05	8.8E−01	4.9E−05	7.4E−07	1.5E−09
2 % CO_2, 16 % H_2O, 82 % Ar	0.04	3.3E−05	9.4E−01	7.0E−05	7.5E−07	1.5E−09
2 % CO_2, 16 % H_2O, 82 % Ar	0.11	3.3E−05	9.3E−01	7.0E−05	7.5E−07	1.5E−09

[a]The viscosity of the gas was calculated using the method of Chung et al. [4]

[b]The density of the gas was determined assuming ideal gas, T = 573 K, P = 1 bar

[c]The molecular diffusion coefficient was obtained using the Fuller et al. [5] equation and Blanc's law [4].

[d]The intraparticle diffusion coefficient was calculated as: $\frac{1}{D_p} = \tau\left(\frac{1}{D_k} + \frac{1}{D_m}\right)$

[e]The value of surface diffusion coefficient presented in this table corresponds to the coefficient at zero coverage, Ds°, estimated by Reijers et al. [6] for CO_2 adsorption on hydrotalcites

Molar composition	m^a (mol CO_2 kg ads^{-1})	b^b (bar^{-1})	H^c (mol CO_2 kg ads^{-1} bar^{-1})	K^d (−)
2 % CO_2, 98 % Ar	0.39	21.70	8.5	492.9
2 % CO_2, 98 % Ar	0.39	21.70	8.5	492.9
2 % CO_2, 16 % H_2O, 82 % Ar	0.52	23.10	12.0	699.5
2 % CO_2, 16 % H_2O, 82 % Ar	0.52	23.10	12.0	699.5

[a]Monolayer capacity in the Langmuir isotherm at 573 K from TGA analysis
[b]Gas-solid interaction parameter in the Langmuir isotherm at 573 K from TGA analysis
[c]Henry's law constant, $H = m \cdot b$
[d]Dimensionless Henry's law constant, $K = K' \cdot \rho_p \cdot T \cdot R$

Adsorbent properties (layered double oxide, Mg/Al ratio = 2)	
Pore diameter[a], d_{pore}	1.04E−08 m
Particle diameter[b], d_p	4.64E−04 m
Knudsen diffusivity[c], D_k	1.82E−06 m^2 s^{-1}
Particle density[d], ρ_p	1222 kg_{ads} m^{-3}
Tortuosity[e], τ	2.35
Particle porosity[f], ε_p	0.56

[a]Determined by N_2 physisorption at 77 K (Sect. 3.4.1)
[b]Measured as described in Sect. 3.4.9
[c]$D_k = 9.7 \times 10^3 \left(\frac{d_{pore}}{2}\right)\left(\frac{T}{M_{CO_2}}\right)^{1/2}$, where d_{pore} is in cm
[d]$\rho_p = \frac{\rho_{skeletal}}{1+\rho_{skeletal} v_{pore}}$, where $\rho_{skeletal}$ was obtained by helium pycnometry (Sect. 3.4.9)
[e]Measured by mercury porosimetry (Sect. 3.4.9)
[f]$\varepsilon_p = v_{pore}\rho_p$, where v_{pore} is the pore volume and was measured by N_2 physisorption

Appendix C
Layered Double Hydroxides Supported on Multi-Walled Carbon Nanotubes for CO_2 Adsorption

C.1 Molecular Formula of the As-Synthesised LDH from EDS/TGA Results (Sect. 4.3.1)

- Proposed composition: $Mg_{0.6}Al_{0.3}(OH)_2(CO_3)_{0.15} \cdot xH_2O$
- Total mass of LDH analysed by TGA: 0.00918 g
- Mass of water lost up to 493 K: 0.00116 g (i.e. 12.7 wt%, 0.000064 mol)
- Mass of dehydrated mineral: 0.00802 g
- Moles of dehydrated mineral: 0.00012 mol (considering 65.67 g mol^{-1})
- Therefore, 1 mol of LDH has = 0.53 mol

C.2 N_2 Isotherm of LDH_1 and LDO_1 (Larger Scale) (Sect. 4.3.3)

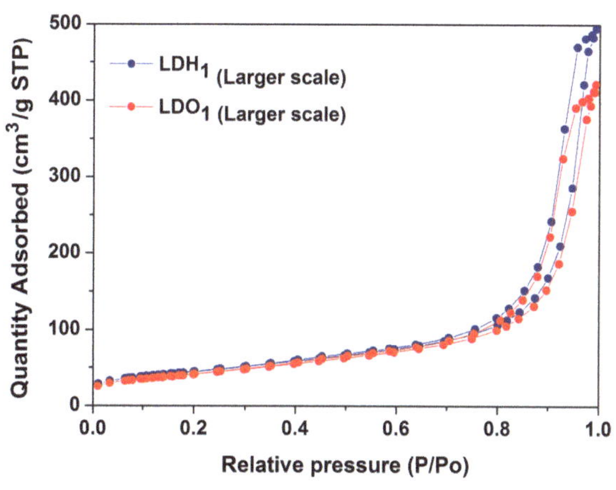

D. Iruretagoyena Ferrer, *Supported Layered Double Hydroxides as CO_2 Adsorbents for Sorption-enhanced H_2 Production*, Springer Theses,
DOI 10.1007/978-3-319-41276-4

C.3 Representative Adsorption Profiles of Pure LDO at Different Temperatures and PCO_2 = 200 mbar (Sect. 4.4)

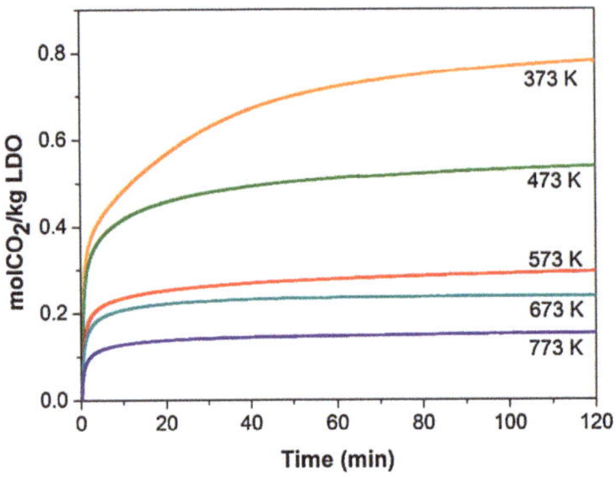

Temperature (K)	CO_2 capacity, 120 min (mol CO_2 kg LDO^{-1})
373	0.78[a]
473	0.54[a]
573	0.29[a]
673	0.24[a]
773	0.15[b]

[a]The adsorbents were activated at 673 K
[b]The adsorbent was activated at 773 K

C.3 Representative Adsorption Profiles of LDO_1 (1:1 LDO/MWCNTs) at Different Temperatures and PCO_2 = 200 mbar (Sect. 4.4)

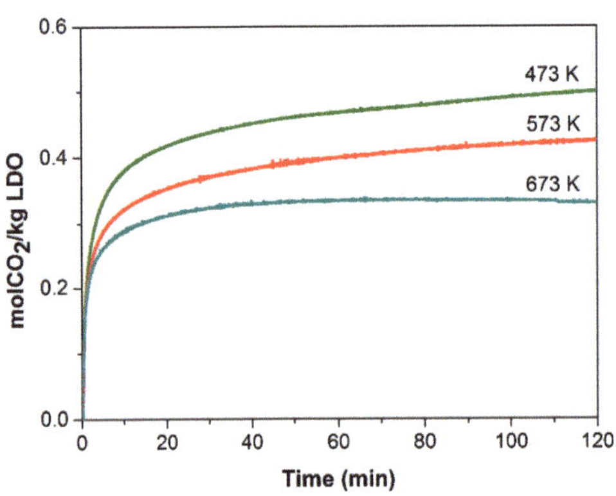

Temperature (K)	CO_2 capacity, 120 min (mol CO_2 kg LDO^{-1})
473	0.50[a]
573	0.42[a]
673	0.33[a]

[a]The adsorbents were activated at 673 K

C.4 XRD Diffraction Pattern of Commercial LDH/α-Alumina (Sect. 4.4)

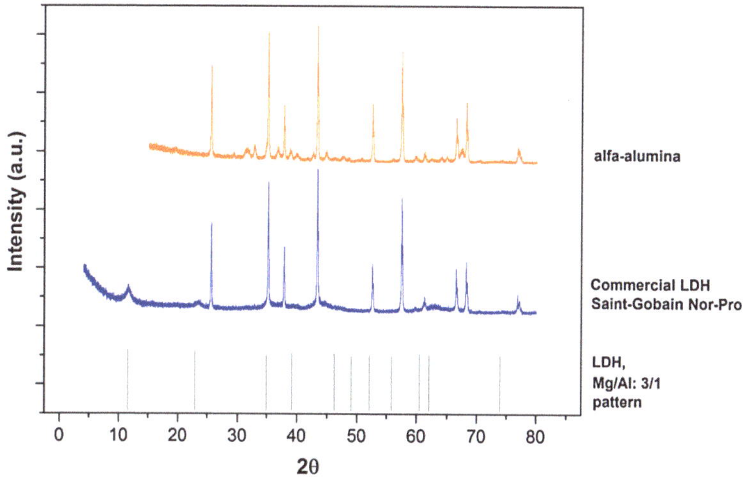

Appendix D
Layered Double Hydroxides Supported on Graphene Oxide for CO_2 Adsorption

D.1 N_2 Isotherm of LDH_{99}, LDH_7 and LDH_3 (Sect. 5.2.3)

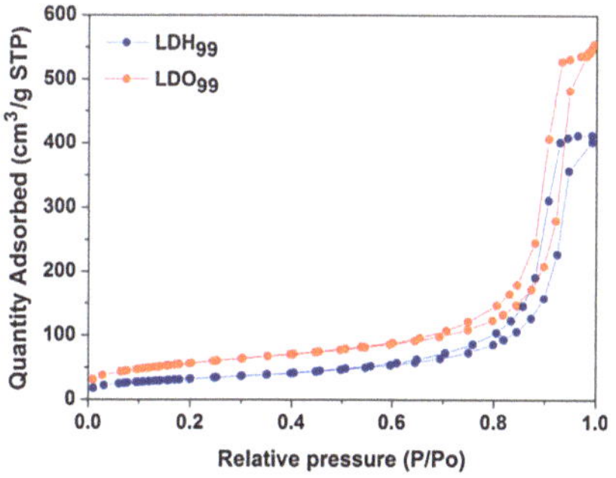

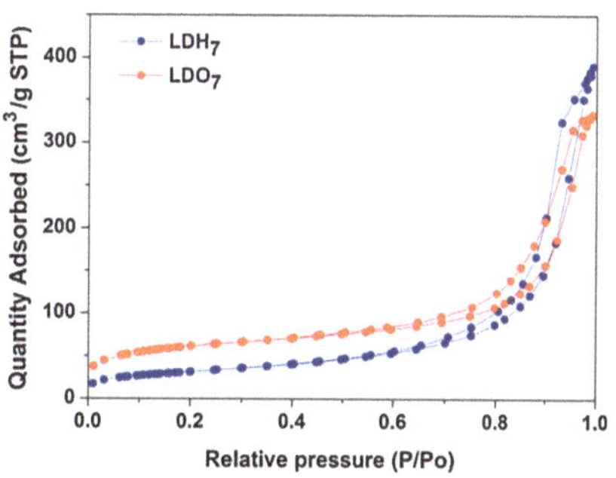

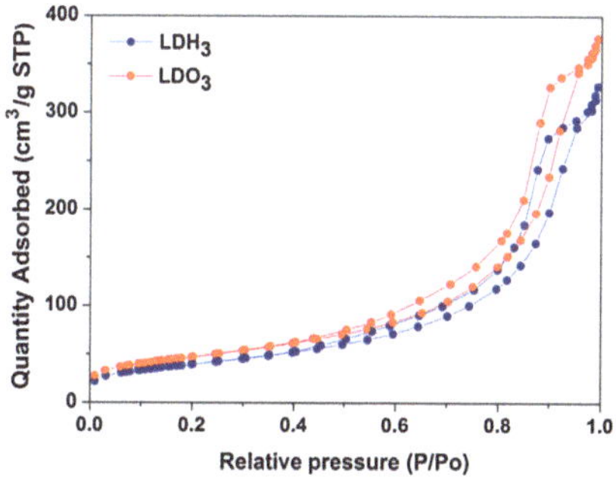

D. Iruretagoyena Ferrer, *Supported Layered Double Hydroxides as CO_2 Adsorbents for Sorption-enhanced H_2 Production*, Springer Theses,
DOI 10.1007/978-3-319-41276-4

D.2 Toth Isotherm Fittings (623 K) for Pure LDO and LDO_{20} (Sect. 5.4.5)

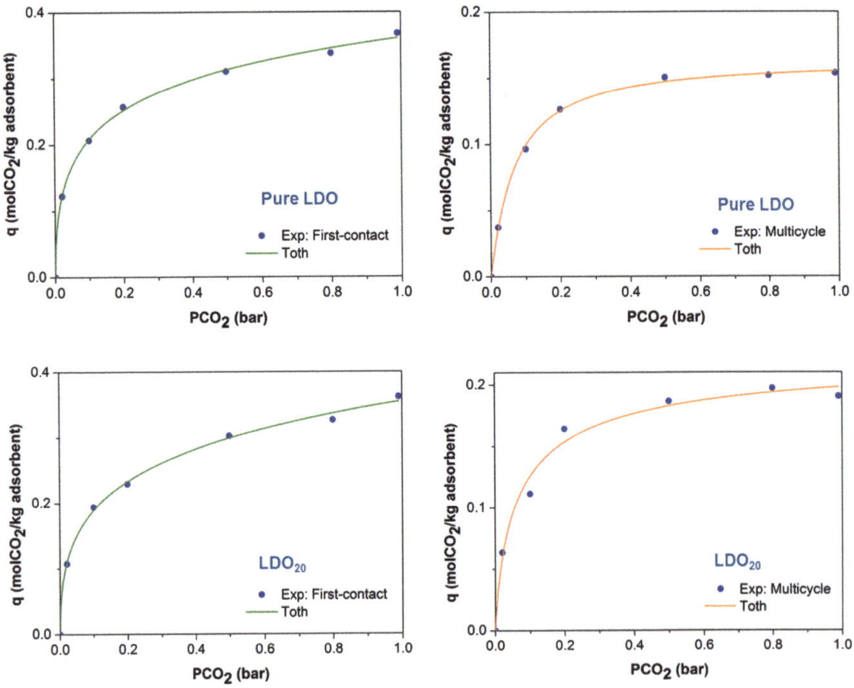

Adsorbent	First-contact: Toth isotherm[a]			
	m (mol kg^{-1})	b (bar^{-1})	*t* (−)	R^2
Pure LDO	0.94 ± 0.43	545 ± 921	0.23 ± 0.08	0.99
LDO_{20}	1.83 ± 2.18	1170 ± 3905	0.17 ± 0.10	0.99

[a]The parameters are shown with the standard errors obtained from the fittings

Adsorbent	Multicycle: Toth isotherm[a]			
	m (mol kg^{-1})	b (bar^{-1})	*t* (−)	R^2
Pure LDO	0.16 ± 0.00	13.0 ± 1.5	1.12 ± 0.12	0.99
LDO_{20}	0.23 ± 0.03	25.5 ± 16.5	0.70 ± 0.28	0.99

[a]The parameters are shown with the standard errors obtained from the fittings

D.3 Composition and Capacities of Adsorbents from Preliminary Studies [7]

Sample name[a]	wt% GO actual	wt% Na[c]	mol CO_2 kg LDO^{-1b}	mol CO_2 kg ads^{-1b}
Pure LDH	0	0.02	0.28 ± 0.01	0.28 ± 0.01
LDH_{20}	7	1.70	0.48 ± 0.02	0.45 ± 0.02
LDH_{10}	12	1.40	0.43 ± 0.08	0.38 ± 0.08
LDH_{1}	33	1.20	0.45 ± 0.07	0.30 ± 0.07
$LDH_{0.1}$	83	1.20	0.57 ± 0.03	0.10 ± 0.03

[a]LDHx, where x is the nominal LDH to GO weight ratio of the sample
[b]The adsorption capacities are shown with standard errors
[c]Measured by ICP

Appendix E
Influence of Alkali Metals on Layered Double Hydroxides Supported on Graphene Oxide for CO_2 Adsorption

E.1 Crystallite Size of the As-Synthesised and Activated Adsorbents (Sect. 6.4)

Sample name[a]	Crystallite size (003) (nm)	Sample name[a]	Crystallite size (200) (nm)
Pure LDH	12	Pure LDO	2.1
$LDH_{I\text{-}Na}$	13	$LDO_{I\text{-}Na}$	2.5
$LDH_{I\text{-}K}$	14	$LDO_{I\text{-}K}$	2.7
$LDH_{I\text{-}Cs}$	16	$LDO_{I\text{-}Cs}$	2.7
LDH_{20}	13	LDO_{20}	2.3
$LDH_{20\text{-}I\text{-}Na}$	15	$LDO_{20\text{-}I\text{-}Na}$	2.7
$LDH_{20\text{-}I\text{-}K}$	15	$LDO_{20\text{-}I\text{-}K}$	2.4
$LDH_{20\text{-}I\text{-}Cs}$	15	$LDO_{20\text{-}I\text{-}Cs}$	2.7

[a]LDHx and LDOx where x is the nominal LDH to GO weight ratio of the sample

E.2 Toth Isotherm Fittings (573 K) for $LDO_{20\text{-}R\text{-}Na}$ (Sect. 6.5)

Toth parameters for multicycle isotherms at 573 K from TGA under dry conditions

Adsorbent	Toth isotherm[a]			
	m (mol kg^{-1})	log b (bar^{-1})	*t* (−)	R^2
$LDO_{20\text{-}R\text{-}Na}$ First-contact	4.1 ± 11.5	9 ± 10.4	0.09 ± 0.13	0.99
$LDO_{20\text{-}R\text{-}Na}$ Multicycle	0.42 ±0.04	1.7 ± 1.5	0.86 ± 0.35	0.99

[a]The parameters are shown with the standard errors obtained from the fittings

D. Iruretagoyena Ferrer, *Supported Layered Double Hydroxides as CO_2 Adsorbents for Sorption-enhanced H_2 Production*, Springer Theses,
DOI 10.1007/978-3-319-41276-4

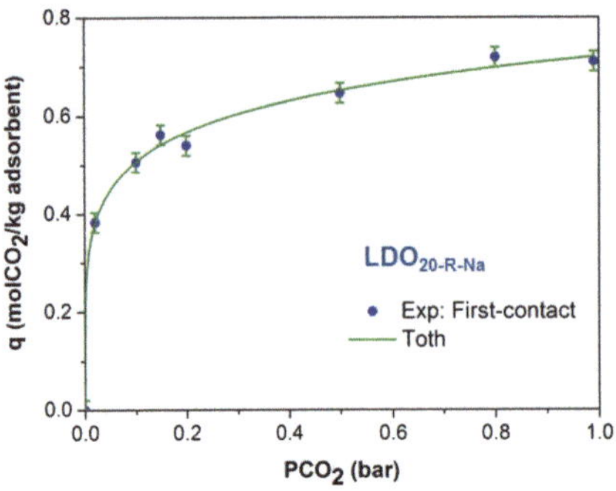

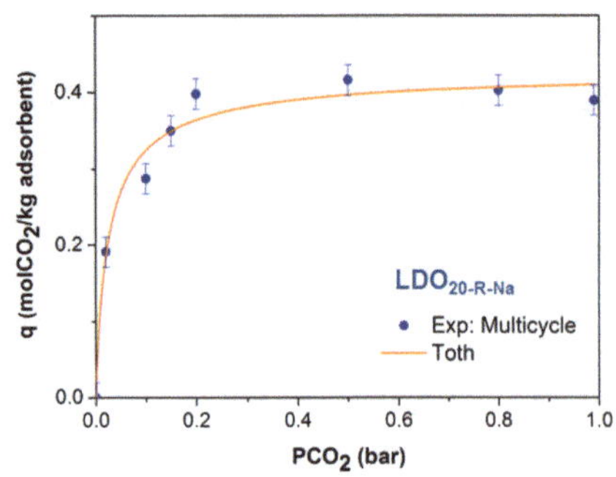

E.3 Langmuir and Freundlich Isotherm Fittings (573 K) (Sect. 6.5)

Parameters for adsorption isotherms at 573 K from TGA under dry conditions

Adsorbent	Langmuir isotherm[a]			Freundlich isotherm[a]		
	m (mol kg^{-1})	b (bar^{-1})	R^2	k (mol kg^{-1})	n (−)	R^2
Multicycle $LDO_{I\text{-}Na}$	0.40 ± 0.01	34.3 ± 4.8	0.99	0.41 ± 0.02	5.9 ± 1.5	0.93
Multicycle $LDO_{20\text{-}I\text{-}Na}$	0.45 ± 0.02	36.8 ± 9.5	0.97	0.47 ± 0.03	6.3 ± 1.7	0.93
Multicycle $LDO_{I\text{-}Cs}$	0.44 ± 0.01	48.7 ± 7.2	0.99	0.46 ± 0.02	7.0 ± 1.4	0.96
Multicycle $LDO_{20\text{-}I\text{-}Cs}$	0.47 ± 0.01	46.6 ± 5.5	0.99	0.48 ±0.02	7.0 ± 1.5	0.96

[a]The parameters are shown with the standard errors obtained from the fittings

E.4 Toth Isotherm Fittings (573 K) (Sect. 6.5)

Toth parameters for multicycle isotherms at 573 K from TGA under dry conditions

Adsorbent	Toth isotherm[a]			
	m (mol kg^{-1})	log b (bar^{-1})	t (−)	R^2
First-contact $LDO_{I\text{-}K}$	1.12 ± 0.33	3.33 ± 3.73	0.26 ± 0.11	0.99
Multicycle $LDO_{I\text{-}K}$	0.52 ± 0.03	1.86 ± 1.66	0.72 ± 0.22	0.99
First-contact $LDO_{20\text{-}I\text{-}K}$	7.4 ± 18.99	0.88 ± 0.11	0.08 ± 0.08	0.99
Multicycle $LDO_{20\text{-}I\text{-}K}$	0.46 ± 0.02	1.7 ± 1.3	0.94 ± 0.26	0.99

[a]The parameters are shown with the standard errors obtained from the fittings

Appendix F
CO_2 Adsorption on Unsupported and Graphene Oxide Supported Layered Double Hydroxides in a Fixed-Bed

F.1 N_2 Purge and Temperature-Swing Cycles Under Dry Conditions (Sect. 7.2.1)

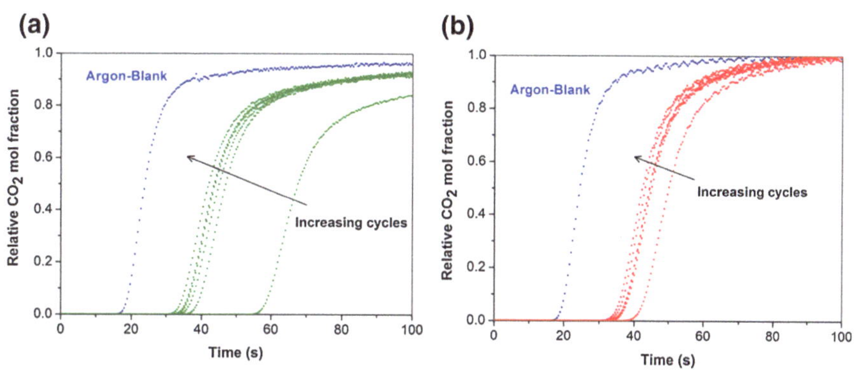

a N_2 purge (six cycles; adsorption 573 K; desorption 573 K) for $LDO_{20\text{-}R\text{-}Na}$ and **b** temperature-swing (six cycles; adsorption 573 K; desorption 673 K) for pure LDO under dry conditions: 1 bar, 10 % CO_2/Ar, 46 ml min^{-1} STP, m_a = 0.3 g

D. Iruretagoyena Ferrer, *Supported Layered Double Hydroxides as CO_2 Adsorbents for Sorption-enhanced H_2 Production*, Springer Theses,
DOI 10.1007/978-3-319-41276-4

F.2 Temperature-Swing Cycles Under Wet Conditions (Sect. 7.2.2)

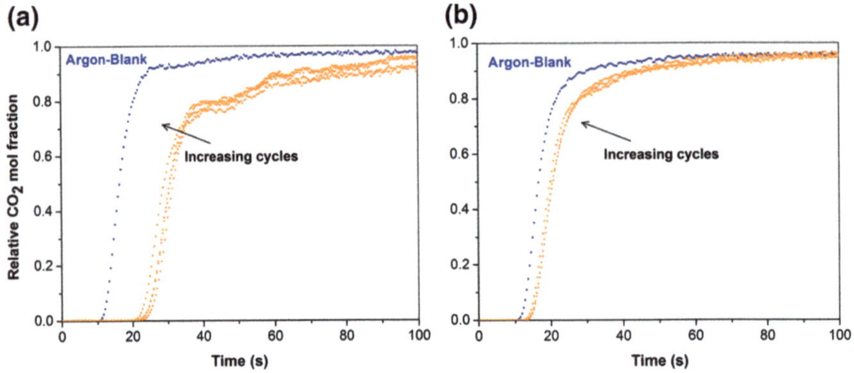

a Temperature-swing (four cycles; adsorption 573 K; desorption 673 K) for LDO_{20} under wet conditions: 1 bar, 1.7 % CO_2/16 % H_2O/Ar, 165 ml min^{-1} STP, $m_a = 0.3$ g. **b** Temperature-swing (three cycles; adsorption 573 K; desorption 673 K) for LDO_{20} under wet conditions: 17 % CO_2/16 % H_2O/Ar, 165 ml min^{-1} STP, $m_a = 0.3$ g

F.3 Temperature Profiles of the Fixed-Bed for the Pure LDO (Sect. 7.2.3)

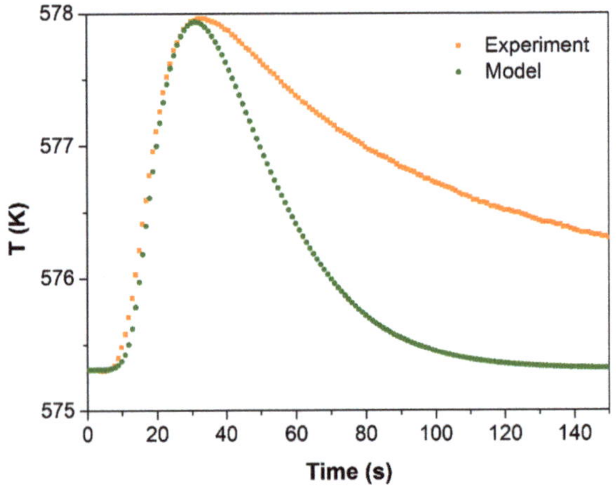

Appendix G
Sorption-Enhanced Methanol-to-Shift for H_2 Production: Thermodynamics and Catalysts Selection

G.1 Equilibrium Calculations of Reactive Systems with Simultaneous CO_2 Adsorption (Sect. 8.2)

A general algorithm to calculate the chemical equilibrium of a reactive system with simultaneous CO_2 adsorption was developed using MATLAB (R2010a). A non-stoichiometric approach based on the direct minimisation of the Gibbs free energy was used to calculate the equilibrium compositions of the system specifying the set of species involved [8, 9, 10].

The total Gibbs free energy of a chemical system at constant temperature and pressure can be expressed as follows:

$$G = \sum_{i=1}^{N} \mu_i n_i, \quad i = 1\ldots N \tag{G.1}$$

To minimise Eq. (G.1), the following constraints are introduced based on the element balances of the system:

$$\sum_{i=1}^{N} n_i a_{ij} = A_j, \quad j = 1\ldots M \tag{G.2}$$

By introducing the Lagrange multipliers, λ_j, a new function $L(n_{i,}\ \lambda_i)$ is defined by Eq. (G.3). The second term in the right hand side of Eq. (G.3) is zero in agreement with Eq. (G.2), and therefore $G(n_i)$ and $L(n_{i,}\ \lambda_i)$ are equal.

$$L(n_i, \lambda_j) = \sum_{i=1}^{N} \mu_i n_i + \sum_{j=1}^{M} \lambda j \left(\sum_{i=1}^{N} n_i a_{ij} - A_j \right) \tag{G.3}$$

To find the critical point of the function $L(n_{i,}\ \lambda_i)$, the derivatives with respect to n_i and λ_i must be zero. It is obvious by simple inspection that this corresponds to a local minimum since the second derivatives of $L(n_{i,}\ \lambda_i)$ are positive. Thus for a

D. Iruretagoyena Ferrer, *Supported Layered Double Hydroxides as CO_2 Adsorbents for Sorption-enhanced H_2 Production*, Springer Theses,
DOI 10.1007/978-3-319-41276-4

gaseous chemical system a set of $N + M$ non-linear algebraic equations can be written as follows:

$$\frac{\partial L(n_i, \lambda_j)}{\partial n_i} = \Delta G_i^f + RT \ln\left(P \frac{n_i}{\sum_{i=1}^{N} n_i}\right) + RT \ln \hat{\phi}_i + \sum_{j=1}^{M} \lambda_j a_{ij} = 0 \qquad \text{(G.4)}$$

$$\frac{\partial L(n_i, \lambda_j)}{\partial \lambda_j} = \left(\sum_{i=1}^{N} n_i a_{ij}\right) - A_j = 0, \qquad \text{(G.5)}$$

Additionally, a particular version of Eq. (G.4) has to be written for carbon dioxide, Eq. (G.6), to account for the fraction f of CO_2 that is being removed by adsorption.

$$\Delta G_{CO_2}^f + RT \ln\left(P \frac{n_{CO_2}(1 - f)}{\sum_{i=1}^{N} n_i - n_{CO_2} f}\right) + RT \ln \hat{\phi}_{CO_2} + \lambda_C + 2\lambda_O = 0 \qquad \text{(G.6)}$$

The degrees of freedom of the problem are zero since there are $N + M$ equations and $N + M$ unknown variables, i.e. the number of moles of species defined and the number of atoms involved represented by the Langrange multipliers. The non-linear algebraic system was solved using the MATLAB in-built routine *fsolve* and the MATLAB code *newtons* [11]. In addition, a subroutine was developed to calculate the fugacity coefficients of each species in the gas mixture using the Redlich-Kwong equation of state.

In order to calculate the equilibrium compositions of a reactive system, the MATLAB program requires the following data to be specified: number of atoms, number of species, temperature, pressure, standard Gibbs free energies [8] and initial guesses of the unknowns. The convergence of some systems was not straightforward and therefore strategies such as scaling up/down the initial guesses and the subsequent rescaling of the solution were applied [11].

The optimisation methodology was validated using some examples from the literature. Thermodynamic data were available in the case of adsorption-enhanced steam reforming of methane [12] and WGS[2] (see Table G.1 and Fig. G.1). In addition, equilibrium data was found for the conventional steam reforming of methanol [13] (see Table G.2).

A good agreement between the results of the MATLAB program and the data from the literature was observed.

Table G.1 Equilibrium composition of sorption-enhanced methane steam reforming (% mol, dry basis)

Species	$f = 0$		$f = 0.999$	
	Literature	Present study	Literature	Present study
CH_4	34	36	4	2
$CO + CO_2$	13	13	<50 ppm	26 ppm
H_2	53	51	96	98

Comparison between the literature [12] and present work. Operating conditions: 723 K, 4.8 bar and S/C = 6

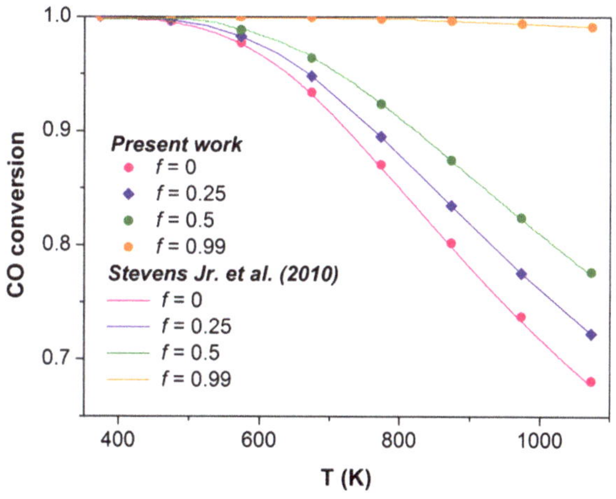

Fig. G.1 CO conversion in WGS: comparison between present work and calculations reported in the literature [2] (operating conditions: 1 bar and S/C = 2)

Table G.2 Equilibrium compositions of methanol steam reforming: comparison between the literature [13] and present work

Equilibrium composition (moles)	Literature	Present study
CH_3OH	0.0003	0.0004
H_2O	0.1364	0.1334
CO	0.1354	0.1330
CO_2	0.8635	0.8666
H_2	2.8620	2.8658

Operating conditions: 500 K, 1 bar and S/C = 1

G.2 Temperature-Programmed RWGS Over Various Catalysts (Sect. 8.3)

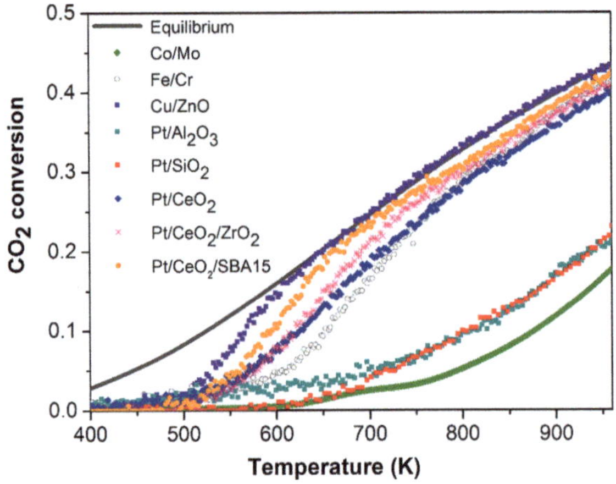

G.3 Temperature-Programmed Methanol Decomposition Over Various Catalysts (Sect. 8.3)

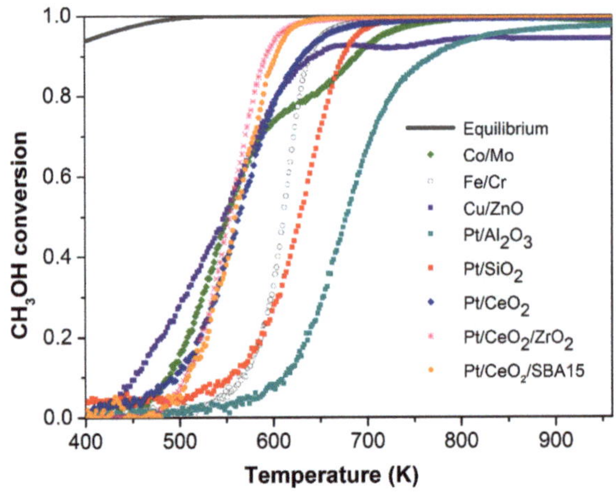

References

1. Halsey, G., & Taylor, H. S. (1947). The adsorption of hydrogen on tungsten powders. *The Journal of Chemical Physics, 15*(9), 624–630.
2. Stevens, R. W, Jr., Shamsi, A., Carpenter, S., & Siriwardane, R. (2010). Sorption-enhanced water gas shift reaction by sodium-promoted calcium oxides. *Fuel, 89*(6), 1280–1286.
3. Vannice, M. A. (2005). *Kinetics of catalytic reactions*. PA, USA: Springer.
4. Poling, B. E., Prausnitz, J. M., & O'Connell, J. P. (2004). *The properties of gases and liquids*. New York, USA: McGraw-Hill.
5. Fuller, E. N., Schettler, P. D., & Giddings, J. C. (1966). A new method for prediction of binary gas-phase diffusion coefficients. *Industrial & Engineering Chemistry, 58*(5), 18–27.
6. Reijers, H. T. J., Boon, J., Elzinga, G. D., Cobden, P. D., Haije, W. G., & van den Brink, R. W. (2009). Modeling study of the sorption-enhanced reaction process for CO_2 capture. I. Model development and validation. *Industrial & Engineering Chemistry Research, 48*(15), 6966–6974.
7. Garcia-Gallastegui, A., Iruretagoyena, D., Gouvea, V., Mokhtar, M., Asiri, A. M., Basahel, S. N., et al. (2012). Graphene oxide as support for layered double hydroxides: Enhancing the CO_2 adsorption capacity. *Chemistry of Materials, 24*(23), 4531–4539.
8. Smith, J. M., Van Ness, H. C., & Abbott, M. M. (2001). *Introduction to chemical engineering thermodynamics*. London, UK: McGraw-Hill.
9. Chen, H., Zhang, T., Dou, B., Dupont, V., Williams, P., Ghadiri, M., & Ding, Y. (2009). Thermodynamic analyses of adsorption-enhanced steam reforming of glycerol for hydrogen production. *International Journal of Hydrogen Energy, 34*(17), 7208–7222.
10. Li, M. (2008). Equilibrium calculation of gaseous reactive systems with simultaneous species adsorption. *Industrial & Engineering Chemistry Research, 47*(23), 9263–9271.
11. Yang, W. Y., Cao, W., Chung, T., & Morris, J. (2005). *Applied numerical methods using MATLAB*. NJ, USA: Wiley.
12. Hufton, J. R., Mayorga, S., & Sircar, S. (1999). Sorption-enhanced reaction process for hydrogen production. *AIChE Journal, 45*(2), 248–256.
13. Lwin, Y., Daud, W. R. W., Mohamad, A. B., & Yaakob, Z. (2000). Hydrogen production from steam–methanol reforming: Thermodynamic analysis. *International Journal of Hydrogen Energy, 25*(1), 47–53.

MIX
Papier aus verantwortungsvollen Quellen
Paper from responsible sources
FSC® C105338

If you have any concerns about our products,
you can contact us on
ProductSafety@springernature.com

In case Publisher is established outside the EU,
the EU authorized representative is:
Springer Nature Customer Service Center GmbH
Europaplatz 3, 69115 Heidelberg, Germany

Printed by Libri Plureos GmbH
in Hamburg, Germany